How can we test if a supermassive black hole lies at the heart of every active galactic nucleus? What are LINERs, BL Lac objects, N galaxies, broad-line radio galaxies, and radio-quiet quasars and how do they compare? This timely textbook answers these questions in a clear, comprehensive, and self-contained introduction to active galactic nuclei – for advanced undergraduates and graduate students in astronomy and physics.

The study of AGNs is one of the most dynamic areas of contemporary astronomy, involving one fifth of all research astronomers. This textbook provides a systematic review of the observed properties of AGNs across the entire electromagnetic spectrum, examines the underlying physics, and shows how the brightest AGNs, quasars, can be used to probe the farthest reaches of the Universe. This book serves as both an entry point to the research literature and as a valuable reference for researchers in the field.

An introduction to active galactic nuclei

AN INTRODUCTION TO
ACTIVE GALACTIC NUCLEI

BRADLEY M. PETERSON

Department of Astronomy, The Ohio State University

CAMBRIDGE
UNIVERSITY PRESS

PUBLISHED BY THE PRESS SYNDICATE OF THE UNIVERSITY OF CAMBRIDGE
The Pitt Building, Trumpington Street, Cambridge, United Kingdom

CAMBRIDGE UNIVERSITY PRESS
The Edinburgh Building, Cambridge CB2 2RU, UK
40 West 20th Street, New York, NY 10011–4211, USA
477 Williamstown Road, Port Melbourne, VIC 3207, Australia
Ruiz de Alarcón 13, 28014 Madrid, Spain
Dock House, The Waterfront, Cape Town 8001, South Africa

http://www.cambridge.org

First published 1997
Reprinted 2001, 2003

Typeset in 10/13pt Times

A catalogue record for this book is available from the British Library

Library of Congress Cataloguing in Publication data
Peterson, B. M. (Bradley M.)
An introduction to active galactic nuclei / Bradley M. Peterson.
   p. cm.
Includes bibliographical references and index.
ISBN 0 521 47348 9. – ISBN 0 521 47911 8 (pbk.)
1. Active galactic nuclei. I. Title.
QB858.3.P47 1997
523.1′12 – dc20 96-26005 CIP

ISBN 0 521 47348 9 hardback
ISBN 0 521 47911 8 paperback

Transferred to digital printing 2004

Contents

Preface

Like many textbooks, this one arose out of the author's frustration. While I believe that there are many excellent journal articles, scholarly reviews, conference proceedings, and even a few advanced monographs on active galactic nuclei (AGNs), there is no single place where a beginning student can get the very basic background necessary to get the most out of the more research-oriented material. The aims of this book are thus actually twofold: first, I wanted to summarize our basic, if marginal, understanding of AGNs at what I believe is a level of familiarity that should be expected of doctoral-level students in astronomy, and second, I wanted to provide a fairly comprehensive introduction to AGNs that would serve as a gateway to the more specialized review articles and research literature for students who have research ambitions in the field. The intended audience is thus advanced undergraduate and beginning graduate students in astronomy and astrophysics. Fairly complete undergraduate preparation in physics is assumed, as is some basic understanding of extragalactic astronomy.

I have tried to focus on basic issues and avoid minutiae and arcane issues, even though some of these undoubtedly will turn out to be tremendously important in the future. I have attempted to compile the basic background material that is by-and-large familiar to researchers in AGNs, although I caution that it is by no means complete: research-level competence in the field of AGNs will require a good deal more background than is given here. As a next step for the serious research student, I would recommend the Saas-Fee course by Blandford, Netzer, and Woltjer (1990), Osterbrock's (1989) text on nebular physics and AGNs, and the more recent review articles (mostly in *Annual Reviews of Astronomy and Astrophysics*) cited in the Bibliography.

The book is obviously written from the point of view of an observer – there is more discussion about the observed properties of AGNs than there is about what physical mechanisms might be operative. Some relevant areas (e.g., radio astronomy and astrophysics, accretion-disk theory, cosmology, nebular physics) receive only a basic introduction here as they are covered in more detail elsewhere.

As will become immediately apparent, our understanding of the AGN phenomenon is still in a primitive state after over 30 years of intensive research. This means that beyond the unproven supermassive black-hole/accretion-disk paradigm, there really is no widely developed consensus about the nature of AGNs. This alone probably explains the dearth of introductory texts, and also serves to make it clear that the content of this book reflects my personal impression about what is essential in the field, and does not necessarily represent a consensus view.

The organization of the book is intended to lead the novice through the basics of the field in a coherent and reasonably self-contained fashion. I have tried to maintain a clear distinction between the things we know, the things we have reason to believe, and the things that we are not really very sure about at all. In the first two chapters, the various types of AGNs and their observed properties are introduced, as much as possible within a historical context to help clarify to the novice some of the basic phenomenology and the traditional distinctions among the various classes of AGNs. Chapters 3–8 provide an introduction to what I believe constitutes the general view of the physical properties of AGNs, with the material organized from the smallest scales to the largest. Following this description of the AGN phenomenon itself, attention is focused on how AGNs can be used to explore the structure and history of the Universe. A brief introduction is provided in Chapter 9 to the basic formalism that describes the structure and evolution of the Universe. While most of this material is certainly covered more completely and authoritatively in any number of books on cosmology, at the very least this chapter introduces the notation that will be used throughout the remainder of the book. The final three chapters discuss how QSOs are isolated in statistically meaningful ways, the luminosity function and space distribution of QSOs, and QSO absorption spectra, with particular attention paid to how these serve as cosmological probes.

In accordance with the usual standards of the astronomical literature, cgs units and appropriate astronomical units are used throughout. I have attempted to keep the notation as consistent as possible with the notation generally used in the literature (the cosmological equations generally follow the convention of Peebles), even though this has sometimes led to some possible ambiguities (e.g., γ is used in various places to refer to the Lorentz factor, the damping constant, and the power-law index describing the redshift dependence of QSO absorption systems), although use of various symbols should be clear in context. One item that needs special attention is that I use h_0 to denote the Hubble constant in units of $100 \ \mathrm{km \, s^{-1} \, Mpc^{-1}}$; the subscript, which is not generally used in the literature, is added to avoid any possible confusion with Planck's constant h.

As this is intended to be a general text and not a research monograph, the direct citations to the literature are incomplete. Indeed many important and outstanding journal articles have not been cited. The references given either justify directly some specific point made in the text, are of historical interest, or just happened to be articles with which I am familiar. I am certain that all astronomers who work on AGNs in the field will feel some consternation at my omission of at least some of the papers that they consider to be particularly important (many, no doubt, that they have authored themselves), and in my own defense I argue only that excessive literature citations would overburden both the reader and the author. I chose to cite only currently published works (no preprints are cited), and whenever possible the citations are to the refereed literature (which has been examined through December 1995) rather than to conference proceedings or hard-to-obtain materials. Some of the material is not attributed to any specific source, usually because the particulars are so well known that they hardly seem to need justification. Much of the literature that shaped my own thinking on AGNs I have cited in the Bibliography. The Bibliography includes not only research review articles, which can supplement the material presented here, but also a list of more-or-less general references on AGNs which I have on one occasion or another found useful or enlightening.

Bradley M. Peterson
Columbus, Ohio

Acknowledgements

One of things that surprised me in the writing of this textbook was how much I learned about AGNs in the process, and not just in parts of the field outside my own research interests. Much of my recent education was acquired through detailed comments on an earlier draft from a number of friends and colleagues. Detailed criticism on all or parts of the text was generously provided by R. A. Edelson, A. V. Filippenko, C. B. Foltz, I. M. George, K. Horne, K. T. Korista, M. A. Malkan, P. L. Martini, P. S. Osmer, R. W. Pogge, B. S. Ryden, M. Santos-Lleó, P. M. Rodríguez-Pascual, J. C. Shields, G. A. Steigman, D. H. Weinberg, and R. J. White. I am especially grateful to each of these individuals for pointing out errors, omissions, misstatements, and unclear or misleading passages, which I hope I have now corrected. Any errors that remain are either due to limitations in what is known about AGNs at the time of writing, or are errors of my own doing, perhaps because I did not listen closely enough to the advice given to me.

Several others were kind enough to enlighten me on some specific issues during the preparation of this book. For these helpful conversations, I wish to thank G. J. Ferland, H. L. Marshall, R. J. Scherrer, I. Wanders, and R. J. Weymann.

Another surprise I had was how much was involved in putting together a useful set of figures and tables. Material for the figures was supplied by T. A. Boroson, J. N. Bregman, A. H. Bridle, G. J. Ferland, A. V. Filippenko, C. B. Foltz, P. J. Francis, I. M. George, P. M. Gondhalekar, R. W. Goodrich, K. T. Korista (who also provided the measurements in Table 1.1), P. S. Osmer, T. J. Pearson, R. W. Pogge, Z. I. Tsvetanov, S. Veilleux, S. J. Warren, D. H. Weinberg, and W. Zheng. I could not possibly have completed this work in a timely fashion without their generous assistance. I also thank S. Mitton and A. Black of the Cambridge University Press who were ready with advice and help when I needed it, and M. Seymour for careful copy-editing.

I am also grateful to the many graduate students at The Ohio State University who suffered through early versions of this text. My thanks go to J. Hunley for checking many of the references. I also wish to thank G. J. Ferland, H. R. Miller, K. Sellgren, and B. J. Wilkes for the kind encouragement they gave me based on the original class notes that led to this book.

Many of the individuals listed above are people that I have worked with closely, some for many years, and some of whom have had a tremendous influence on how I think about AGNs, although I absolve them from any blame for the contents of this book. I particularly wish to thank my many past and present faculty colleagues at The Ohio State University for their support (and tolerance) over the years, notably J. A.

Baldwin, G. J. Ferland, G. H. Newsom, P. S. Osmer, R. W. Pogge, and especially E. R. Capriotti, for whom words alone will not suffice. R. W. Pogge (who provided much assistance with the figures) and P. S. Osmer were extraordinarily helpful and patient during the final stages of writing this book; hardly a day passed without me cornering them to discuss one thing or another, and never once did they let on how they must have felt about these constant intrusions.

Finally, I wish to thank my family not only for tolerance and forbearance, but for their constant encouragement, especially during some difficult times. I dedicate this book to my wife, Jan, and our children, who cannot really understand why anyone would want to own a copy of a book like this, and to my parents, Harry and Dona, who cannot understand why anyone would not.

1 Basic Properties and a Brief Historical Perspective

In general the term 'active galactic nucleus', or AGN, refers to the existence of energetic phenomena in the nuclei, or central regions, of galaxies which cannot be attributed clearly and directly to stars. The two largest subclasses of AGNs are Seyfert galaxies and quasars, and the distinction between them is to some degree a matter of semantics. The fundamental difference between these two subclasses is in the amount of radiation emitted by the compact central source; in the case of a typical Seyfert galaxy, the total energy emitted by the nuclear source at visible wavelengths is comparable to the energy emitted by all of the stars in the galaxy (i.e., $\sim 10^{11} L_\odot$), but in a typical quasar the nuclear source is brighter than the stars by a factor of 100 or more. Historically, the early failure to realize that Seyferts and quasars are probably related has to do with the different methods by which these two types of objects were first isolated, which left a large gap in luminosity between them. The appearance of quasars did not initially suggest identification with galaxies, which is a consequence of the basic fact that high-luminosity objects, like bright quasars, are rare. One is likely to find rare objects only at great distances, which is of course what happens with quasars. At very large distances, only the star-like nuclear source is seen in a quasar, and the light from the surrounding galaxy, because of its small angular size and relative faintness, is lost in the glare of the nucleus. Hence, the source looks 'quasi-stellar'.

1.1 Seyfert Galaxies

The first optical spectrum of an active galaxy was obtained at Lick Observatory by E. A. Fath in 1908 as part of his dissertation work. He noted the presence of strong emission lines in the nebula NGC 1068. V. M. Slipher at Lowell Observatory obtained a higher-quality, higher-resolution spectrum of NGC 1068, and commented that the emission lines are similar to those seen in planetary nebulae. He also made the important observation that these lines are resolved, and have widths of hundreds of kilometers per second.

Carl Seyfert (1943) was the first to realize that there are several similar galaxies which form a distinct class. Seyfert selected a group of galaxies on the basis of high central surface brightness, i.e., stellar-appearing cores. Seyfert obtained spectra of these galaxies and found that the optical spectra of several of these galaxies (NGC 1068,

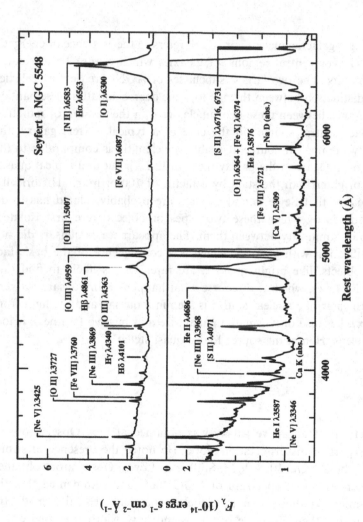

Fig. 1.1. The optical spectrum of the Seyfert 1 galaxy NGC 5548. The prominent broad and narrow emission lines are labeled, as are strong absorption features of the host galaxy spectrum. The vertical scale is expanded in the lower panel to show the weaker features. The full width at half maximum (FWHM) of the broad components is about 5900 km s^{-1}, and the width of the narrow components is about 400 km s^{-1}. The strong rise shortward of 4000 Å is the long-wavelength end of the 'small blue bump' feature which is a blend of Balmer continuum and Fe II line emission. This spectrum is the mean of several observations made during 1993 with the 3-m Shane Telescope and Kast spectrograph at the Lick Observatory. Data courtesy of A. V. Filippenko.

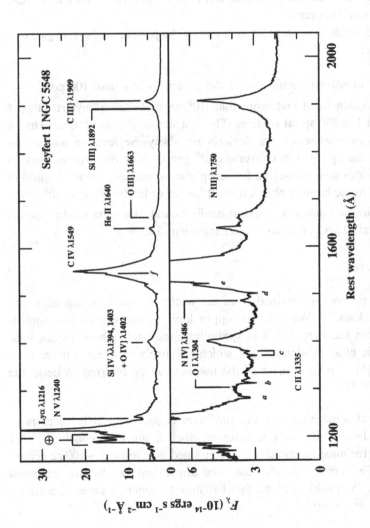

Fig. 1.2. The ultraviolet spectrum of the Seyfert 1 galaxy NGC 5548. The prominent broad emission lines are labeled. The emission labeled with the Earth symbol ('⊕') arises in the extended upper atmosphere of the Earth and is known as 'geocoronal' emission. Most of the labeled absorption features arise in our own Galaxy and thus appear blueshifted from their rest wavelengths since the spectrum has been corrected for the redshift of NGC 5548 ($z = 0.017$). The labeled absorption features are O II $\lambda 1302$ (a), C II $\lambda 1335$ (b), Si IV $\lambda\lambda 1394$, 1403 (c), Si II $\lambda 1527$ (d), and C IV $\lambda\lambda 1548$, 1551 (e). Another weak C IV $\lambda\lambda 1548$, 1551 doublet (f) is only slightly displaced shortward of line center and presumably arises in NGC 5548 itself. This spectrum is the mean of several observations obtained with the Faint Object Spectrograph on the *Hubble Space Telescope* in 1993. Data courtesy of K. T. Korista.

NGC 1275,† NGC 3516, NGC 4051, NGC 4151, and NGC 7469) are dominated by high-excitation nuclear emission lines (Figs. 1.1 and 1.2). The important characteristics of these spectra were found to be:

(1) The lines are broad (up to 8500 km s^{-1}, full width at zero intensity).
(2) The hydrogen lines sometimes are broader than the other lines.

Seyfert galaxies received no further attention until 1955, when NGC 1068 and NGC 1275 were detected as radio sources.

Woltjer (1959) made the first attempt to understand the physics of Seyfert galaxies. He noted the following:

(1) The nuclei are unresolved, so the size of the nucleus is less than 100 pc.
(2) The nuclear emission must last more than 10^8 years, because Seyfert galaxies constitute about 1 in 100 spiral galaxies. This is a simple argument. One extreme scenario is that galaxies which are Seyferts are always Seyferts, in which case their lifetime is the age of the Universe (10^{10} years). The opposite extreme is one where all spirals pass through a Seyfert phase (or phases) – since 1 spiral in 100 is currently in the Seyfert phase, it must last of order $10^{10}/100 = 10^8$ years.
(3) If the material in the nucleus is gravitationally bound, the mass of the nucleus must be very high. This is a simple virial argument, i.e.,

$$M \approx v^2 r/G. \tag{1.1}$$

The velocity dispersion is obtained from the widths of the emission lines and is of order 10^3 km s^{-1}. We have an upper limit to the size of the nucleus ($r \lesssim 100$ pc) from the fact that it is spatially unresolved. The emission lines are characteristic of a low-density gas, which effectively provides a lower limit $r \gtrsim 1$ pc (eq. 6.15). Thus the mass of the nucleus can be inferred to be in the range $M \approx 10^{9\pm1}\ M_\odot$.

Point (iii) tells us that something very extraordinary is occurring at the centers of Seyfert galaxies. If a large value of r is assumed, then it must be concluded that something like 10% of the mass of the galaxy is contained in a volume ~ 100 pc across. On the other hand, if r is much smaller than the upper limit set by ground-based spatial resolution, then the problem to be faced is how to generate an extraordinary amount of energy in a tiny volume.

---

† Minkowski (1957) pointed out that NGC 1275 is atypical of the rest of the Seyfert class because it shows extended emission at two radial velocities separated by $\Delta v \approx 3000$ km s^{-1}. The lower-velocity emission lines are associated with the central galaxy of the Perseus cluster, which emits a Seyfert-type spectrum and also shows extended nebulosity that may originate in a 'cooling flow' from the hot intracluster medium. The higher-velocity emission lines appear to arise in a star-forming cluster member that lies along our line of sight to NGC 1275 itself.

1.2   Radio Surveys and Quasars

Quasars were originally discovered as a result of the first radio surveys of the sky in the late 1950s. By this time, the angular resolution of radio observations was good enough to identify the strongest radio sources with individual optical objects, often galaxies, but sometimes stellar-appearing sources. Important early surveys included the following:

3C and 3CR: The third Cambridge (3C) catalog (Edge *et al.* 1959), based on observations at 158 MHz, and its revision, the 3CR catalog (Bennett 1961), at 178 MHz, detected sources down to a limiting flux of 9 Jy.† The 3C catalog is not limited by flux sensitivity, but by 'confusion' of sources at low flux levels; fainter sources become so numerous on the sky that they cannot be unambiguously distinguished from one another with poor angular resolution observations. There are 471 3C sources, and 328 3CR sources, numbered sequentially by right ascension (e.g., 3C 273, which is at epoch 1950.0 coordinates $\alpha_{1950} = 12^{\mathrm{h}} 26^{\mathrm{m}} 33^{\mathrm{s}}.35$, $\delta_{1950} = +02° 19' 42''$), with the numbering between the two catalogs kept as consistent as possible. Sources which appear in the 3CR but not the 3C are kept in proper right ascension order by appending a decimal point and additional digit to the immediately preceding source (e.g., 3C 390.3). All 3C sources are north of $-22°$ declination, but the 3CR excludes sources south of $-5°$.

PKS: This was an extensive survey (Ekers 1969) of the southern sky (declination $< +25°$) undertaken at Parkes (PKS), Australia, originally at 408 MHz (detection limit 4 Jy) and later at 1410 MHz (to 1 Jy) and 2650 MHz (to 0.3 Jy). These sources are designated by 1950.0 position, using the format 'HHMM±DDT', where HHMM refers to the hours and minutes of right ascension (epoch 1950.0), ± is the sign of the declination, and DDT is the declination, in degrees (DD) and tenths (T) of degree‡ (e.g., 3C 273 = PKS 1226+023); this is still the most common, and useful, system of naming quasars.

4C: The fourth Cambridge survey was a more sensitive version (limiting flux 2 Jy) of the 3C, again undertaken at 178 MHz (Pilkington and Scott 1965, Gower, Scott, and Wills 1967). Names of 4C sources are given as '±DD.NN', where ±DD is the source declination, and NN is a sequence number within the declination band (e.g., 3C 273 = 4C 02.32).

AO: The Arecibo Occultation (AO) survey is notable for the extremely accurate positions it produced as a result of observing radio sources as they were occulted by

---

† A jansky (Jy), named after the pioneering radio astronomer Karl Jansky, is a unit of specific flux, and in early radio astronomy was simply called a 'flux unit'. It is defined as $1 \text{ Jy} = 10^{-26} \text{ watts m}^{-2} \text{ Hz}^{-1} = 10^{-23} \text{ ergs s}^{-1} \text{ cm}^{-2} \text{ Hz}^{-1}$.

‡ The tenths of degree designation was added only after the number of radio sources became rather large. Since the 1980s, some catalogs have begun using declination tags of ±DDMM, where MM is in arcminutes.

the Moon (Hazard, Gulkis, and Bray 1967). As a small angular size radio source is occulted by the Moon, the moment of disappearance identifies its position as being somewhere along the locus of points defined by the preceding limb of the Moon. Subsequent occultations of the same source give additional such loci, all of which intersect at a single point. Thus, the location of the points is limited by timing accuracy and the accuracy to which the position of the limb of the Moon is known, not by the size of the radio beam. Names of these sources are Parkes-style.

Ohio: The Ohio radio survey (e.g., Ehman, Dixon, and Kraus 1970) was made with a unique 79×21 m transit telescope at 1415 MHz. The unusual geometry of the telescope produces an irregular beam of half-power beam width $\sim 10'$ in right ascension and $\sim 40'$ in declination. Source names are given as 'Ox-NNN', where x is a letter indicating an hour-wide band of right ascension (with 'A' and 'O' excluded†, so sources between 0^h and 1^h are 'OB' and sources between 23^h and 0^h are 'OZ') and NNN is a serial number (e.g., 3C 273 = ON 044). As in many of the early surveys, the positions and fluxes are not reliable (particularly at sub-jansky flux levels), but the Ohio survey is notable in that up through the late 1970s some of the most distant and most luminous known quasars (e.g., OQ 172 = 1442+101 and OH 471 = 0642+449) were Ohio sources.

Most radio sources at high Galactic latitude were identified with resolved galaxies. However, the positions of some of these radio sources were found to be coincident with objects that looked like stars on normal photographs, such as the Palomar Sky Survey. The first strong radio source unambiguously identified with a star-like optical source was 3C 48. On the basis of a radio position obtained with a two-element interferometer, Matthews and Sandage (1963) found that the optical counterpart of this source was a magnitude 16 star. However, the photographic spectra obtained of this source were very confusing, as they showed strong very broad emission lines at unidentified wavelengths. It is worth noting how unsuitable photographic spectra are for work on quasars; it was unclear to the first investigators whether these broad features were emission lines or merely the continuum between broad absorption lines, as in white dwarf spectra. Photometry of these objects revealed that they are anomalously blue (relative to normal stars). Needless to say, the nature of these 'radio stars' was very uncertain.

Another radio source identified with a star-like object, in this case on the basis of an accurate lunar occultation measurement (Hazard, Mackey, and Shimmins 1963), was 3C 273. The first breakthrough in understanding these extraordinary objects came with Maarten Schmidt's realization (Schmidt 1963) that the emission lines seen in the spectrum of this source were actually the hydrogen Balmer-series emission lines and Mg II $\lambda 2798$ at the uncommonly large redshift $z = 0.158$, where we recall that the

† The 'OA' sources actually constitute a preliminary version of the Ohio catalog which was compiled by Kraus (1966).

redshift is an observational quantity defined by the observed wavelength λ of a spectral line relative to its laboratory wavelength λ_0,

$$z = \frac{\lambda - \lambda_0}{\lambda_0} = \frac{\lambda}{\lambda_0} - 1. \tag{1.2}$$

This redshift was approximately an order of magnitude larger than those of the original Seyfert galaxies and was among the largest ever measured at the time, with only a few very faint rich clusters of galaxies rivaling it. The obvious interpretation of the redshift is that it is of cosmological origin, a consequence of the expansion of the Universe, and the Hubble law thus gives the distance

$$d = cz/H_0 = 3000zh_0^{-1} \text{ Mpc} = 470h_0^{-1} \text{ Mpc}, \tag{1.3}$$

where h_0 is the Hubble constant in units of $100 \text{ km s}^{-1} \text{ Mpc}^{-1}$. More disturbing than this vast distance was the enormous luminosity implied; 3C 273 was and remains the brightest known quasar ($B = 13.1$ mag). Using the formula for the distance modulus

$$m - M = 5\log(cz/H_0) + 25 \tag{1.4}$$

(where cz/H_0 is measured in Mpc†), the absolute magnitude of 3C 273 is $M_B = -25.3 + 5\log h_0$, which is about 100 times as luminous as normal bright spirals like the Milky Way or M31. Once the redshift mystery was unlocked, identification of lines in 3C 48 (Greenstein and Matthews 1963) and other quasar spectra followed quickly.

As the physical nature of these luminous star-like objects was not understood, they became known simply as 'quasi-stellar radio sources', a term which was subsequently shortened to 'quasars'.‡

The probable importance of quasars was recognized immediately. The extremely high luminosities of these objects implied physical extremes that were not found elsewhere in the nearby Universe. The suggestion that massive black holes might be involved appeared early (e.g., Zel'dovich and Novikov 1964). The possible role of active nuclei in galaxy formation and evolution was also seen (e.g., Burbidge, Burbidge, and Sandage 1963). The high luminosities of quasars also imply that they might serve as important cosmological probes, since they could be in principle detected and identified at very large distances. These considerations provided early and continuing strong motivation for finding quasars and studying their properties. As the number of known quasars increased, progressively greater and greater redshifts were identified, and as of the mid-1990s the highest observed redshifts are $z \approx 5$.

---

† We will see in Chapter 9 that this formula for the distance is in general an approximation that is valid for only small values of z.

‡ The term 'quasar' is attributed to H.-Y. Chiu (1964). It was some time, however, before the word was generally accepted into the astronomy lexicon. Indeed, under the editorial direction of S. Chandrasekhar, *The Astrophysical Journal* resisted use of 'quasar' for many years, finally conceding with 'regrets' in 1970 (see Schmidt 1970). An informal vote on the term at the Second Texas Symposium on Relativistic Astrophysics in 1964 yielded 20 ayes and 400 or so abstentions (Robinson, Schild, and Schucking 1965).

1.3   Properties of Quasars

By late 1964, Schmidt had studied a sufficient number of quasars to define their properties (Schmidt 1969):

- Star-like objects identified with radio sources.
- Time-variable continuum flux.
- Large UV flux.
- Broad emission lines.
- Large redshifts.

Not all objects that we now call AGNs share every one of these properties. In fact, the single most common characteristic of AGNs is probably that they are all luminous X-ray sources (Elvis *et al.* 1978). Nevertheless, the characteristics identified by Schmidt are important for understanding the physics of AGNs and as well as understanding the techniques by which AGNs are found. We will therefore discuss these various properties individually below.

In more modern terms, one of the defining characteristics of quasars is their very broad spectral energy distribution, or SED (Fig. 1.3). Quasars are among the most luminous objects in the sky at every wavelength at which they have been observed. Unlike spectra of stars or galaxies, AGN spectra cannot be described in terms of blackbody emission at a single temperature, or as a composite over a small range in temperature. Non-thermal processes, primarily incoherent synchrotron radiation, were thus invoked early on to explain quasar spectra.

In general, the broad-band SED of a quasar continuum can be characterized crudely as a power law

$$F_\nu = C\nu^{-\alpha}, \tag{1.5}$$

where α is the power-law index, C is a constant, and F_ν is the specific flux (i.e., per unit frequency interval), usually measured in units of $\mathrm{ergs\, s^{-1}\, cm^{-2}\, Hz^{-1}}$. The convention that we adopt throughout this book is that a positive spectral index characterizes a source whose flux density *decreases* with increasing frequency. The reader is cautioned that some authors absorb the minus sign into the definition of the spectral index (e.g., $F_\nu = C\nu^\alpha$). Specific fluxes, particularly in the ultraviolet and optical parts of the spectrum, are often measured per unit wavelength interval (i.e., in units of $\mathrm{ergs\, s^{-1}\, cm^{-2}\, Å^{-1}}$) rather than per unit frequency interval. The total flux measured in any bandpass is the same, of course, whether the bandwidth is measured in frequency or in wavelength, so the relationship $F_\nu d\nu = F_\lambda d\lambda$ always holds. The transformation between the two systems is thus

$$F_\nu = F_\lambda \left| \frac{d\lambda}{d\nu} \right| = \frac{\lambda^2 F_\lambda}{c}, \tag{1.6}$$

so an equivalent form to eq. (1.5) is $F_\lambda = C'\lambda^{\alpha-2}$.

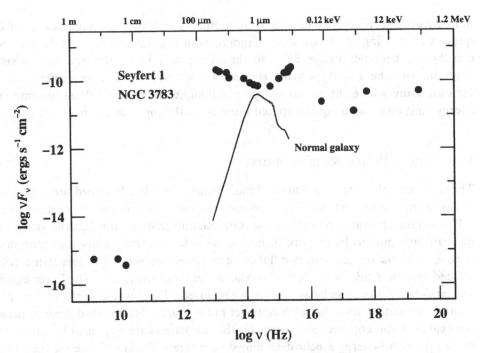

Fig. 1.3. The spectral energy distribution (SED) of the Seyfert 1 galaxy NGC 3783 (Alloin *et al.* 1995), from radio to γ-ray energies. Shown for comparison is SED for a normal (type Sbc) galaxy, from a template spectrum compiled by Elvis *et al.* (1994). The flux scale of the normal galaxy spectrum has been adjusted to give the correct relative contribution of AGN component and starlight for NGC 3783 (in mid-1992) at 5125 Å through a 5″ × 10″ spectrograph aperture.

Fits to quasar spectra over large frequency ranges yield spectral indices that are typically in the range $0 \lesssim \alpha \lesssim 1$, but different values of α are found for different spectral ranges. Indeed, it is obvious that a single power law cannot hold over all frequencies as the integrated power would diverge either at high frequencies (for $\alpha \leq 1$) or at low frequencies (for $\alpha \geq 1$). Over a frequency range $v_1 - v_2$, the total power received is

$$P(v_1, v_2) = \int_{v_1}^{v_2} F_v \, dv = C \int_{v_1}^{v_2} v^{-\alpha} \, dv$$
$$= \frac{C}{1 - \alpha} \left(v_2^{1-\alpha} - v_1^{1-\alpha} \right) \quad (\alpha \neq 1)$$
$$= C \ln \left(\frac{v_2}{v_1} \right) \quad (\alpha = 1). \tag{1.7}$$

The case $\alpha = 0$, a 'flat spectrum' on a conventional plot of specific flux versus frequency, has equal energy per unit frequency interval and the case $\alpha = 1$ has equal energy per unit *logarithmic* frequency interval.† A useful way to plot the broad-band SED in quasars is on a $\log \nu F_\nu$ versus $\log \nu$ diagram, as in Fig. 1.3. In this case, the power-law distribution becomes $\nu F_\nu \propto \nu^{1-\alpha}$, so the case $\alpha = 1$ is a horizontal line in such a diagram, and the $\alpha = 0$ spectrum rises with frequency. This is the preferred format for examining where the quasar energy is actually emitted as it reflects the amount of energy emitted in each equally spaced interval on the logarithmic frequency axis.

1.3.1   Radio Properties of Quasars

The radio morphology of quasars and radio galaxies is often described broadly in terms of two components, 'extended' (i.e., spatially resolved) and 'compact' (i.e., unresolved at $\sim 1''$ resolution), that have different spectral characteristics, although the synchrotron mechanism seems to be at work in both cases. The extended-component morphology is generally double, i.e., with two 'lobes' of radio emission more or less symmetrically located on either side of the optical quasar or center of the galaxy. The linear extent of the extended sources can be as large as megaparsecs. The position of the optical quasar is often coincident with that of a compact radio source. The major difference between the extended and compact components is that the extended component is optically thin to its own radio-energy synchrotron emission, whereas this is not true for the compact sources.

Although a detailed discussion of synchrotron radiation is beyond the scope of this book, we will summarize some of the basic properties of synchrotron-emitting sources. For a homogeneous source with constant magnetic field B, a power-law continuum spectrum (eq. 1.5) can be generated by the synchrotron mechanism by an initial power-law distribution of electron energies E of the form

$$N(E)\,dE = N_0\,E^{-s}\,dE, \tag{1.8}$$

where $\alpha = (s-1)/2$. For the extended component, a typical observed value of the power-law index is $\langle \alpha \rangle \approx 0.7$, so $\langle s \rangle \approx 2.4$. This applies at higher frequencies where synchrotron self-absorption is not important. At lower frequencies, the emitting gas is optically thick and the the trend towards increasing flux with decreasing frequency turns over to yield $F_\nu \propto \nu^{5/2}$. The turn-over frequency increases with the density of relativistic electrons in the source, although it depends on other parameters as well. The relativistic particle densities in the extended radio components are low enough that they are optically thin even at very low radio frequencies (at least to the 3C frequency of 158 MHz). Radio spectra sometimes curve downward at higher frequencies (i.e., α increases with ν). The basic reason for this is that electrons radiate at frequencies proportional to their energies E, and the *rate* at which they lose energy is proportional to E^2. Thus, the highest-energy electrons radiate away their energy the most rapidly,

---

† The case $\alpha = 1$ is sometimes described as having equal energy per decade (or octave), i.e., over any factor of 10 (or two) in frequency.

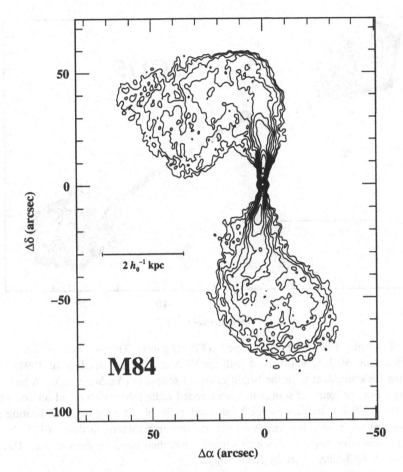

Fig. 1.4. Example of an Fanaroff–Riley type I (FR I) galaxy. This is a radio map of the Virgo cluster elliptical galaxy M84 (3C 272.1 = NGC 4373), based on 4.9 GHz data obtained with the VLA by Laing and Bridle (1987). Data courtesy of A. H. Bridle, figure by R. W. Pogge.

thus depleting the emitted spectrum at the high-frequency end first if no replenishment of the high-energy electrons occurs.

Extended radio structures can be divided into two separate luminosity classes (Fanaroff and Riley 1974). Class I (FR I) sources are weaker radio sources which are brightest in the center, with decreasing surface brightness towards the edges. In contrast, the more luminous FR II sources are limb-brightened, and often show regions of enhanced emission either at the edge of the radio structure or embedded within the structure. Bridle and Perley (1984) give $L_\nu(1.4\,\mathrm{GHz}) = 10^{32}$ ergs s^{-1} Hz^{-1} as the transition specific luminosity between the two types. Quasars are FR II sources. Examples of FR I and FR II sources are shown in Figs. 1.4 and 1.5, respectively.

The characteristics of compact sources are quite different from those of the extended

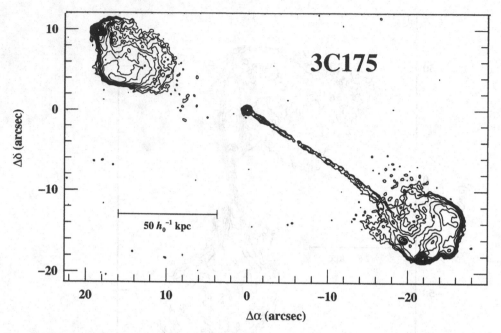

Fig. 1.5. Example of Fanaroff–Riley type II (FR II) galaxy. This is a map of the $z = 0.768$ quasar 3C 175, as observed with the VLA at 4.9 GHz (Bridle *et al.* 1994). The quasar itself is coincident with the bright compact source at $(\Delta\alpha, \Delta\delta) = (0,0)$. A jet extending from the compact source to the extended radio lobes is observed on one side of the source. FR II sources are edge-brightened, probably because of shock heating as the radio-emitting plasma interacts with the ambient intergalactic medium. FR I sources are not edge-brightened, which suggests that their outflows are subsonic. Data courtesy of A. H. Bridle, figure by R. W. Pogge.

sources. Very long-baseline interferometry (VLBI) yields upper limits on the sizes of compact sources typically no better than ~ 0.01 pc. In contrast to the steep spectra of extended sources, the radio spectra of compact sources are usually flat, i.e., with $\alpha \leq 0.5$, if a power-law form is assumed. Unless the initial electron energy distribution in the compact source is quite flat and quite different from that which drives the extended sources, a synchrotron spectrum is only approximately flat near the turnover frequency and even then only over a limited frequency range. The flatness of compact source spectra over several orders of magnitude in frequency is thus usually attributed to either source inhomogeneity or the presence of a number of unresolved small discrete sources within the compact core. In either case, different parts of the compact region become optically thick at different frequencies, which can flatten the integrated spectrum of the compact source over a suitably broad frequency range.

In addition to the compact and extended components, radio sources also often have features known as 'jets',† which are extended linear structures (Bridle and Perley 1984). An example of a jet is seen in the FR II source shown in Fig. 1.4. Jets appear to originate at the central compact source and lead out to the extended lobes. They often show bends or wiggles between the central source and the point where the jet appears to expand into the extended radio structure. The appearance of jets suggests that they transport energy and particles from the compact source to the extended regions. Jets often appear on only one side of the radio source, and in cases where jets are seen on both sides one side (the 'counter-jet') is much fainter than the other. The difference in brightness is thought to be primarily attributable to 'Doppler beaming' which preferentially enhances the surface brightness on the side that is approaching the observer.

The relative strength of the extended, compact, and jet components varies with frequency since the different components have different spectral shapes. The relative strengths also show considerable variation from source to source, with 'lobe-dominated' sources having steep spectra and 'core-dominated' sources having flat spectra. At least part of the observed differences among quasars must be due to orientation effects; whereas the extended components probably emit their radiation isotropically, the compact and jet components emit anisotropically. This will be discussed further in Chapters 4 and 7.

Because of the different observing frequencies and detection limits of the various radio surveys used to find quasars, each of the major radio-source catalogs contains certain selection biases. For example, the 3C sources tend to be those which are the very brightest at low frequencies, which will bias selection toward the high-luminosity, steep-spectrum (and thus lobe-dominated) sources. The 4C survey was more sensitive and tended to turn up sources which are in general of lower luminosity. On the other hand, the Parkes survey, which was undertaken at higher frequencies, tended to select a relatively greater fraction of flat-spectrum (and thus core-dominated) sources. Steep-spectrum sources, on the other hand, can be missed because they can fall below the detection limit at higher frequencies. For example, a source near the 4C detection limit (2 Jy at 178 MHz) will fall below the Parkes detection threshold (1 Jy at 1410 MHz) for $\alpha \gtrsim 0.34$.

1.3.2 Variability

Quasars are variable in every waveband in which they have been studied, not only in the continuum but in the broad emission lines as well. Optical continuum variability of quasars was established even before the emission-line redshifts were understood (e.g, Matthews and Sandage 1963), and variability was one of the first properties of quasars to be explored in detail (e.g., Smith and Hoffleit 1963). A few quasars had been

---

† The term 'jet' was first used in extragalactic astronomy by Baade and Minkowski (1954) to describe the optical linear feature in M87 that was first mentioned by Curtis (1913). Despite the implication of the name, there is no unambiguous evidence that jets involve high outflow velocities, except in the innermost regions (§4.4.2).

identified as unusual 'variable stars' (e.g., GQ Comae = 1202+281 and BL Lacertae = 2200+420) before their spectral properties were known. Many quasars were found to be variable at the 0.3–0.5 mag level over time scales of a few months. A few sources were found to vary significantly on time scales as short as a few days. One can thus conclude on the basis of coherence arguments that much of the radiation must come from a region of order light days (1 light day = 2.54×10^{15} cm) in size. This was immediately perceived as a major problem, since a nucleus comparable in size to the Solar System is emitting hundreds of times as much energy as an entire galaxy.

1.3.3 Ultraviolet Fluxes

Quasars are often found to have unusually blue broad-band optical colors. Johnson *UBV* photometry in particular shows that $U-B$ is remarkably small (i.e., large negative values), and this is often referred to as the 'ultraviolet (UV) excess' of quasars. Quasars occupy a region on the two-color diagram which is not heavily populated by stars, as shown in Fig. 1.6.

It must always be kept in mind that statements such as 'quasars are very blue objects' or 'quasars show an ultraviolet excess' are relative statements, which refer to their spectral energy distribution relative to stars. One must remember that in most stars there is relatively little flux in the U band; in cooler stars, the U band is in the Wien tail of the blackbody distribution, and in hotter stars, the Balmer continuum absorption edge occurs in the center of the band (i.e., at 3646 Å), so there is a real deficit in the number of photons in the shortward half of the band. Any AGN-type power-law spectrum shows a higher ratio of U flux to B flux than does an A star, even though there is really less energy per unit frequency at shorter wavelengths than at longer. Thus, the fundamental reason that quasars appear to have an ultraviolet excess is that quasar spectra are flatter than A-star spectra through the U and B bands.

1.3.4 Broad Emission Lines

The UV–optical spectra of quasars are distinguished by strong, broad emission lines. The strongest observed lines are the hydrogen Balmer-series lines (Hα λ6563, Hβ λ4861, and Hγ λ4340), hydrogen Lyα λ1216, and prominent lines of abundant ions (Mg II λ2798, C III] λ1909, and C IV λ1549); these lines appear in virtually all quasar spectra, but depending on the redshift of the quasar, some may not be observable if they fall outside the spectral window of a particular detector. Typical flux ratios are given in Table 1.1, along with typical equivalent widths, defined by

$$W_\lambda = \int \frac{F_\mathrm{l}(\lambda) - F_\mathrm{c}(\lambda)}{F_\mathrm{c}(\lambda)} d\lambda, \tag{1.9}$$

where $F_\mathrm{l}(\lambda)$ is the observed flux across the emission line at the wavelength λ, and $F_\mathrm{c}(\lambda)$ is the continuum level underneath the emission line. Both of these quantities are specific fluxes that are conventionally measured in units of ergs s^{-1} cm^{-2} Å^{-1}, so the equivalent

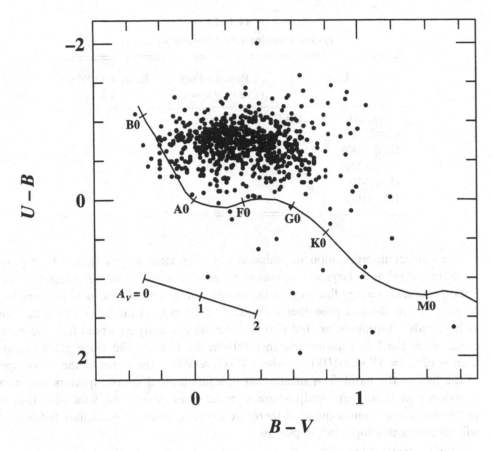

Fig. 1.6. The locations of 788 quasars from the Hewitt and Burbidge (1993) catalog on a two-color ($U - B$ vs. $B - V$) diagram. The locus of the zero-age main sequence, with spectral types indicated, is also shown. The line in the lower left shows how points will be translated due to amounts of reddening corresponding to visual extinction A_V (based on the extinction curve of Cardelli, Clayton, and Mathis (1989) with $R_V = A(V)/E(B - V) = 3.1$). Most quasars with measured UBV magnitudes are color-selected, i.e., identified as quasar candidates on the basis of their blue colors, especially $U - B < 0$.

width is measured in ångströms.† It is nearly always sufficient to approximate eq. (1.9) by

$$W_\lambda = \frac{F_{\text{line}}}{F_c(\lambda)}, \qquad (1.10)$$

where F_{line} is the total line flux, usually in units of ergs s^{-1} cm^{-2}.

---

† X-ray fluxes are sometimes given in units like photons s^{-1} cm^{-2} keV^{-1}, so equivalent widths of X-ray lines are given in keV.

Table 1.1

Typical Emission-Line Strengths in AGNs

Line	Relative Flux (Lyα + N v = 100)	Equivalent Width (Å)
Lyα λ1216 + N v λ1240	100	75
C IV λ1549	40	35
C III] λ1909	20	20
Mg II λ2798	20	30
Hγ λ4340	4	30
Hβ λ4861	8	60

The correct interpretation of emission-line equivalent widths is that they provide an estimate of how large a continuum range one would need to integrate over to obtain the same energy flux as is in the emission line. This is especially relevant in the context of broad-band photometry of quasars; for example, if we compare the colors of high-redshift quasars, we find that the U band is greatly enhanced for quasars with $z \approx 2$, since the Lyα emission line then falls in the U band and alone attributes more than $\sim 12\%$ ($\approx W_\lambda(\text{Ly}\alpha)/W(U)$, where $W(U) \approx 680$ Å, the width of the U bandpass) of the flux in the band. This means that in a flux-limited sample, quasars with strong emission lines in a given bandpass are more likely to be detected, which can thus lead to the erroneous conclusion that there are excess quasars at particular redshifts; we will return to this topic in Chapter 10.

1.3.5   Quasar Redshifts

The first few quasars discovered had redshifts that were comparable to those of the most distant known clusters of galaxies. As more and more quasars were discovered with the refinement of techniques for isolating them (see the next section and Chapter 10), the maximum measured redshifts continued to increase dramatically. By the mid-1970s, several quasars with $z \gtrsim 3$ had been found. The distributions of known quasar redshifts and apparent magnitudes as of 1993 are shown in Figs. 1.7 and 1.8, respectively. Aside from interest in how these sources produce such copious radiation over a broad spectral range, it was also recognized that quasars provide a possibly unique probe of the early Universe – the light that we are now detecting from the most distant known quasars was emitted by them when the Universe was only a small fraction of its current age, and has been in transit since. Quasars are still the only discrete objects that can be observed with relative ease at $z \gtrsim 1$, and thus they are a potentially important cosmological probe. However, in the context of cosmological studies, quasars must be used judiciously. For example, early attempts at producing a Hubble diagram for quasars (as in Fig. 1.9) were not very enlightening because the

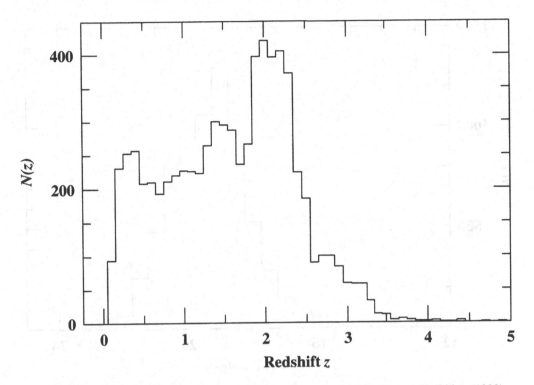

Fig. 1.7. The redshift distribution of 7236 quasars in the Hewitt and Burbidge (1993) catalog.

luminosity function for quasars is very broad, and evolves with time (there are more luminous quasars at high redshift; Chapter 11). An important early finding was that the number of quasars per unit volume reaches a maximum somewhere around $z \approx 2$, even after correction for the Lyα selection effect mentioned in §1.3.4; at earlier epochs (i.e., higher redshifts), they are comparatively rare. Detection of very high-redshift quasars remains of great interest because their existence provides an important constraint on the formation of large structures in the early Universe as well as on the formation of heavy elements, which are clearly seen in the spectra of all quasars.

By the late 1960s, it was also apparent from high-resolution, high signal-to-noise ratio spectra that quasars often have absorption lines in addition to the strong emission features (Chapter 12). The absorption lines are generally much narrower than the emission lines, and are usually detected at redshift *lower* than the emission-line redshift of the quasar itself. These absorption features are thought for the most part to arise in material unassociated with the quasars which lies at lower cosmological redshifts. Thus, quasars also provide an important probe as luminous background sources against which otherwise possibly undetectable structures can be observed.

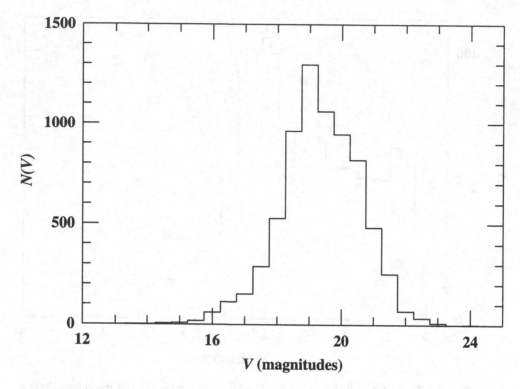

Fig. 1.8. The distribution in apparent *V* magnitude of 7110 quasars in the Hewitt and Burbidge (1993) catalog.

1.4   Radio-Quiet Quasars

It was very quickly realized that the laborious task of examining the fields of radio sources was not the only way to find quasars. Any of the characteristics discussed above could be used to isolate quasar candidates. Ryle and Sandage (1964) noted that a search for *U*-excess objects would be particularly simple. The simplest approach is to take two photographs of a portion of the sky, one photograph with a *B* filter and one with a *U* filter. The relative exposure times are selected so that stars of spectral type A have equal intensities on the two photographs. Comparison of the photographs in a blink comparator leads to the immediate identification of all the *U*-excess objects, which form the minority of objects, as these have brighter *U* images than *B* images, unlike all of the cooler stars. At high Galactic latitude, contamination of the derived sample by O and B stars should be negligible.

This technique was employed very effectively by Sandage on the Mt. Wilson 100-inch telescope and by Lynds with the KPNO 84-inch telescope; in fact, the technique worked so well that many more UV-excess objects were found than suspected on the basis of the number of the surface density of known radio-selected quasars. Many of the first quasars selected by color were drawn from surveys for faint blue objects like

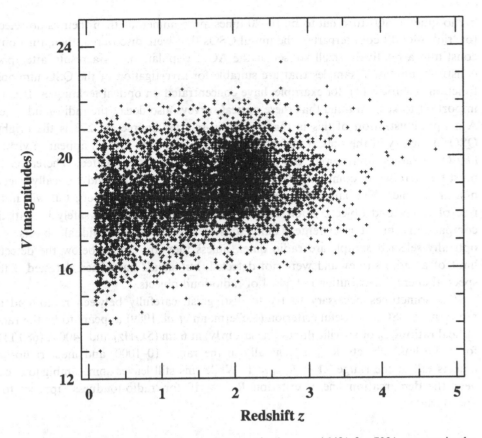

Fig. 1.9. The Hubble diagram (apparent magnitude vs. redshift) for 7031 quasars in the Hewitt and Burbidge (1993) catalog. The vertical width of the distribution is due to the broad quasar luminosity function, i.e., at any redshift, the wide range of apparent magnitudes is due primarily to the wide range of quasar absolute magnitudes.

white dwarfs and horizontal-branch stars such as the PHL ('Palomar–Haro–Luyten'; Haro and Luyten 1962) catalog. Sandage (1965), on the basis of further data obtained on the very wide-field (6° × 6°) Palomar 48-inch Schmidt telescope, estimated that the total density of 'blue stellar objects' was as high as ~3 per square degree down to magnitude 18.5. After some initial confusion, it was realized that most of these high-latitude sources were in fact white dwarfs and RR Lyrae stars; nevertheless, there was a substantial population of quasar-like objects which could be optically selected. These were dubbed 'quasi-stellar objects', or QSOs. The terms 'quasar' and 'QSO' are, however, now used virtually interchangeably, except by purists, who still reserve the term 'quasar' for radio-loud quasi-stellar sources.

The long-wavelength SEDs of optically selected QSOs turn out to be markedly different from those of radio-selected quasars in that the radio emission (relative to the UV–optical–IR) is typically around 100 times lower in the radio. These so-called 'radio-quiet' QSOs turn out to be 10–20 times more numerous than their radio-selected (or 'radio-loud') counterparts. The initial QSOs that were discovered, then, turn out to constitute a relatively small subset of the AGN population. As a result, attempts to construct 'unbiased' samples that are suitable for investigation of the QSO luminosity function (Chapter 11), for example, have concentrated on optical techniques. It is thus important to keep in mind that these optical surveys also detect the radio-loud objects. A simple illustration of this is that the radio-selected source 3C 273 is the brightest QSO in the sky in the optical and other wavebands as well – it will appear in virtually *any* QSO sample that covers the part of the sky where it is located. Therefore, care must be exercised in comparing the properties of 'radio-selected' and 'optically selected' quasar samples. The latter will be dominated by radio-quiet objects, but will not be free of radio-loud objects. Such samples sometimes may be used safely in statistical comparisons, but it is extremely dangerous to assume that individual objects in an optically selected sample are radio quiet, unless they are clearly below the detection limit of a radio survey and were not detected, but *should* have been detected if their spectral energy distribution is typical of radio-loud objects.

It is sometimes necessary to try to distinguish carefully between radio-loud and radio-quiet QSOs. A useful criterion (Kellermann *et al.* 1989) appears to be the radio–optical ratio R_{r-o} of specific fluxes (in, say, mJy) at 6 cm (5 GHz) and 4400 Å (680 THz); for radio-loud objects R_{r-o} is generally in the range 10–1000, and most radio-quiet objects fall in the range $0.1 < R_{r-o} < 1$. While this still leaves some ambiguous cases near the demarcation line, a criterion $R_{r-o} \geq 10$ for 'radio-loudness' appears to be appropriate.†

---

† This criterion means that on a $\log \nu F_\nu$ diagram, the radio spectrum of a radio-loud object can fall as much as four decades below the optical spectrum (see Fig. 4.1).

2   Taxonomy of Active Galactic Nuclei

The taxonomy of AGNs tends to be rather confusing as we do not yet understand the physics underlying the AGN phenomenon. Undoubtedly some of the differences we see between various types of AGNs are due more to the *way* we observe them than to fundamental differences between the various types; this is a theme that will be revisited in Chapter 7. We will introduce the various types of AGNs that are generally recognized, and try to make clear as we proceed how these various types may or may not be related.

2.1   Seyfert Galaxies

Seyfert galaxies are lower-luminosity AGNs, with $M_B > -21.5 + 5 \log h_0$ for the active nucleus the generally accepted criterion, due originally to Schmidt and Green (1983), for distinguishing Seyfert galaxies from quasars. A Seyfert galaxy has a quasar-like nucleus, but the host galaxy is clearly detectable. The original definition of the class (Seyfert 1943) was primarily morphological, i.e., these are galaxies with high surface brightness nuclei, and subsequent spectroscopy revealed unusual emission-line characteristics. Observed directly through a large telescope, a Seyfert galaxy looks like a normal distant spiral galaxy with a star superimposed on the center. The definition has evolved so that Seyfert galaxies are now identified spectroscopically by the presence of strong, high-ionization emission lines. Morphological studies indicate that most if not all Seyferts occur in spiral galaxies (Chapter 8).

Khachikian and Weedman (1974) were the first to realize that there are two distinct subclasses of Seyfert galaxies which are distinguished by the presence or absence of broad bases on the permitted emission lines. Type 1 Seyfert galaxies have two sets of emission lines, superposed on one another. One set of lines is characteristic of a low-density (electron density $n_e \approx 10^3$–10^6 cm^{-3}) ionized gas with widths corresponding to velocities of several hundred kilometers per second (i.e., somewhat broader than emission lines in non-AGNs), and are referred to as the 'narrow lines' (Chapter 6). A second set of 'broad lines' (Chapter 5) are also seen, but in the permitted lines only. These lines have widths of up to 10^4 km s^{-1}; the absence of broad forbidden-line emission indicates that the broad-line gas is of high density ($n_e \approx 10^9$ cm^{-3} or higher) so the non-electric-dipole transitions are collisionally suppressed. Type 2 Seyfert galaxies differ from Seyfert 1 galaxies in that only the narrow lines are present in type 2 spectra.

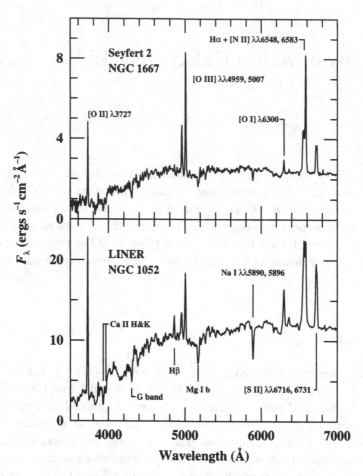

Fig. 2.1. The optical spectra of the Seyfert 2 galaxy NGC 1667 (top) and the LINER (see §2.4) NGC 1052 (bottom) are shown, with important emission lines identified (Ho, Filippenko, and Sargent 1993). Some strong absorption lines that arise in the host galaxy rather than the AGN itself are also identified. These spectra can be compared with the Seyfert 1 spectrum shown in Fig. 1.1, and the QSO spectrum shown in Fig. 2.2. Important differences between Seyfert 2s and LINERs are apparent: the [O III] $\lambda5007$/Hβ flux ratio is much larger in Seyfert 2s (in NGC 1667, the weak Hβ line is obscured by blending with the stellar Hβ *absorption* line) than in LINERs, and low-ionization lines ([N II] $\lambda\lambda6716$, 6731, [S II] $\lambda\lambda6548$, 6583, [O II] $\lambda3727$, and [O I] $\lambda6300$) are all relatively prominent in LINER spectra. Data courtesy of A. V. Filippenko.

A sample Seyfert 2 optical spectrum is shown in the top panel of Fig. 2.1, and this can be compared with the Seyfert 1 spectrum shown in Fig. 1.1.

In addition to the strong emission lines, weak absorption lines due the late-type giant stars in the host galaxy are also observed in both type 1 and type 2 Seyfert spectra; the absorption lines are relatively weak because the starlight is diluted by the non-stellar 'featureless continuum'.† Indeed, the AGN continuum is usually so weak in Seyfert 2 galaxies that it is very difficult to isolate it from the stellar continuum unambiguously.

The narrow-line spectra are clearly distinguishable from the H II-region spectra seen in some normal galaxies, as the Seyfert spectra show a wide range in ionization level, which is typical of a gas ionized by a source where the input continuum spectrum falls off slowly (relative to a Wien law) at ionizing wavelengths. A common, but sometimes misleading (see §2.4), spectroscopic criterion for distinguishing Seyfert galaxies from H II-region galaxies is that the flux ratio [O III] $\lambda 5007/\text{H}\beta > 3$.

The origin of the differences between Seyferts of type 1 and type 2 is not known. There are a few clear examples, to be discussed in Chapter 7, where galaxies have been identified as type 2 Seyferts because the broad components of the lines have proven to be very hard to detect. One school of thought holds that all Seyfert 2s are intrinsically Seyfert 1s where we are unable to see the broad components of the lines from our particular vantage point. It is not clear, however, that this hypothesis can explain all of the observed differences between the two subclasses.

Osterbrock (1981) has introduced the notation Seyfert 1.5, 1.8, and 1.9, where the subclasses are based purely on the appearance of the optical spectrum, with numerically larger subclasses having weaker broad-line components relative to the narrow lines. In Seyfert 1.9 galaxies, for example, the broad component is detected only in the Hα line, and not in the higher-order Balmer lines. In Seyfert 1.8 galaxies, the broad components are very weak, but detectable at Hβ as well as Hα. In Seyfert 1.5 galaxies, the strengths of the broad and narrow components in Hβ are comparable. Caution must be exercised in using this subclassification of Seyfert 1 spectra; whereas the statistics of occurrence might conceivably have some bearing on unification issues (Chapter 7), there are some cases in which broad emission-line variability is so pronounced that the subclassification changes with time. Indeed, there have been cases reported where the broad lines in Seyfert 1 galaxies have nearly completely disappeared (e.g., Penston and Pérez 1984) when the nucleus has faded to a very faint state. Based on casual inspection, the source would have been classified as a Seyfert 2 galaxy. However, close examination of the spectra seems to indicate that the broad lines never *completely* disappear.

2.2 Quasars

Quasars comprise the most luminous subclass of AGNs, with nuclear magnitudes $M_B < -21.5 + 5 \log h_0$. As explained in Chapter 1, a small minority (~ 5–10%) of these sources are the strong radio sources that originally defined the quasar class. Quasars are distinguished from Seyfert galaxies in that in general they are spatially unresolved on

---

† The term 'featureless continuum' means without the emission features and without absorption features due to stars. This term came into fairly wide use as a replacement for the early descriptor 'non-thermal continuum'. Both of these phrases refer specifically to the AGN component of the spectrum as opposed to the starlight, but neither description is accurate.

the Palomar Sky Survey photographs, which means in practice that they have angular sizes smaller than $\sim 7''$. Many of these sources, however, are surrounded by a low surface brightness halo (sometimes called 'quasar fuzz'), which does indeed appear to be starlight from the host galaxy, and a few sources have other peculiar morphological features, such as optical jets (e.g., 3C 273). Quasar spectra are remarkably similar to those of Seyfert galaxies, except that (a) stellar absorption features are very weak, if detectable at all, and (b) the narrow lines are generally weaker relative to the broad lines than is the case in Seyfert galaxies. A 'typical' QSO spectrum, constructed by averaging observations of a large number of QSOs, is shown in Fig. 2.2.

2.3   Radio Galaxies

Strong radio sources are typically identified with giant elliptical galaxies, although some of the brightest radio sources are associated with quasars. Two types of radio galaxies have optical spectra of the sort that we identify with AGN activity; broad-line radio galaxies (BLRGs) and narrow-line radio galaxies (NLRGs) are the radio-loud analogs of type 1 and type 2 Seyfert galaxies, respectively. As a class, they have a number of differences from their radio-quiet counterparts, but in terms of basic phenomenology, they can be considered to be radio-loud Seyferts. One important difference is that they appear to occur in elliptical galaxies rather than spirals.

2.4   LINERs

A very low nuclear-luminosity class of low-ionization nuclear emission-line region galaxies (LINERs) was identified by Heckman (1980). Spectroscopically, they resemble Seyfert 2 galaxies, except that the low-ionization lines, e.g., [O I] $\lambda 6300$ and [N II] $\lambda\lambda 6548, 6583$, are relatively strong. LINERs are very common, and might be present at detectable levels in nearly half of all spiral galaxies (Ho, Filippenko, and Sargent 1994). A sample LINER spectrum is shown in the bottom panel of Fig. 2.1.

As mentioned earlier (§2.1), the [O III]/Hβ flux ratio is often used to distinguish Seyfert galaxies from other types of emission-line galaxies. The criterion that the flux ratio [O III]/Hβ > 3 in AGNs is not a robust indicator, however, because this flux ratio is also typical of low-metallicity H II regions. Indeed, LINER, Seyfert-galaxy, and H II-region spectra cannot be unambiguously distinguished from one other on the basis of any single flux ratio from any pair of lines. However, Baldwin, Phillips, and Terlevich (1981) have shown that various types of objects with superficially similar emission-line spectra (i.e., characteristic of a 10^4 K gas) can be distinguished by considering the intensity ratios of *two* pairs of lines; the relative strengths of various lines are a function of the *shape* of the ionizing continuum, and they therefore can be used to distinguish between, for example, blackbody and power-law ionizing spectra. Figure 2.3 is an example of a 'BPT' (for Baldwin, Phillips, and Terlevich) diagram which

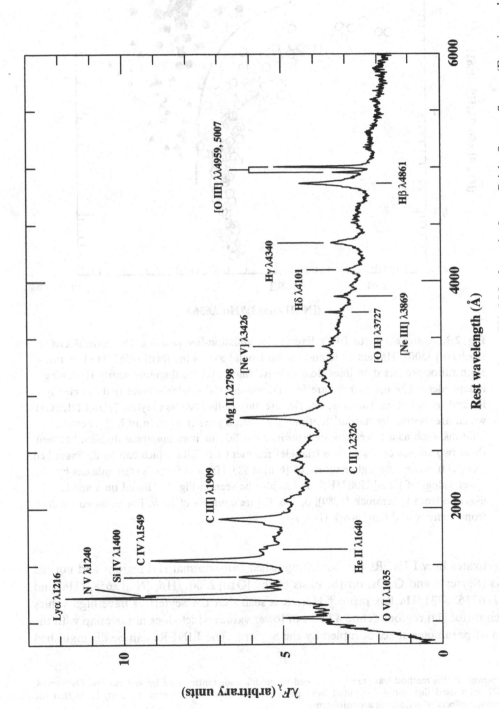

Fig. 2.2. A mean QSO spectrum formed by averaging spectra of over 700 QSOs from the Large Bright Quasar Survey (Francis *et al.* 1991). Prominent emission lines are indicated. Data courtesy of P.J. Francis and C. B. Foltz.

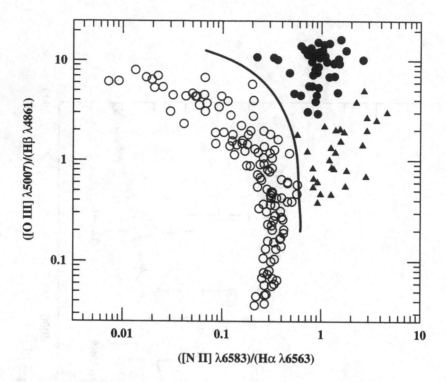

Fig. 2.3. A diagnostic (or BPT) diagram for emission-line galaxies. The vertical axis is the [O III] $\lambda5007$/Hβ flux ratio and the horizontal axis is the [N II] $\lambda6583$/Hα flux ratio. Both ratios are based on lines close in wavelength and are therefore nearly reddening independent. The open circles are for H II regions and similar sources that are clearly ionized by hot stars. The closed circles are narrow-line AGNs (Seyfert 2s and NLRGs) which are ionized by 'harder' continua (i.e., with a greater fraction of high-energy photons, such as a power-law spectrum); the solid line is an empirical division between these two classes of object. The triangles represent LINERs, which can be distinguished from H II regions by higher values of [N II] $\lambda6583$/Hα, and from Seyfert galaxies by lower values of [O III] $\lambda5007$/Hβ, as can also be seen in Fig. 2.1. Based on a similar diagram from Osterbrock (1989), p. 346. Figure courtesy of R. W. Pogge, based on data from Veilleux and Osterbrock (1987).

demonstrates how LINERs can be distinguished from normal H II regions and normal AGNs (Seyferts and QSOs) on the basis of the [O III] $\lambda5007$/Hβ, [N II] $\lambda6583$/Hα, and [S II] $\lambda\lambda6716$, 6731/Hα flux ratios.† Here it is seen that the Seyfert 2s have high values of each ratio. H II regions define a locus of lower values which does not overlap with the region of parameter space occupied by the Seyferts. The LINERs can be distinguished

---

† The power of this method was greatly enhanced by modifications introduced by Veilleux and Osterbrock (1987), who used flux ratios computed from pairs of lines which are close in wavelength, so that the systematic effects of reddening are minimized.

from the Seyfert 2s by their low values of [O III] $\lambda 5007$/Hβ relative to [N II] $\lambda 6583$/Hα, and from the H II regions by their larger values of [N II] $\lambda 6583$/Hα.

Some models indicate that the emission-line spectra of LINERs are consistent with photoionization by a Seyfert-like continuum which is very dilute. The presence of strong [O I] $\lambda 6300$ is especially indicative of a power-law ionizing spectrum, because the ionization potential of O^0 is nearly identical to that of H^0; the [O I] line, which is collisionally excited, will only occur in a zone which has a sufficiently high electron density and temperature to excite the upper level. With a stellar input spectrum, these conditions only occur within the H^+ Strömgren sphere, where the O^0 abundance is negligible. However, a gas ionized by a relatively flat power-law spectrum has an extended partially ionized zone where the [O I] emission arises.

The relationship between LINERs and AGNs is not completely clear. Some, but by no means all, LINERs appear to be simply very low-luminosity Seyfert galaxies. LINER-type spectra can also be produced in cooling flows, in starburst-driven winds, and in shock-heated gas (Heckman 1987, Filippenko 1992).

2.5 BL Lac Objects and Optically Violent Variables

In general, AGNs show continuum variability at all wavelengths at which they have been observed, from X-rays to radio wavelengths. A small subset of AGNs show short time-scale variations that are abnormally large, e.g. $\Delta m \gtrsim 0.1$ mag in the visible wavelengths on time scales as short as a day. In addition to their large variations in flux, they also tend to have high polarization (up to a few percent, as opposed to less than $\sim 1\%$ for most AGNs) which also varies, in both magnitude and position angle, and these sources are always radio loud. These AGNs are known as 'optically violent variables', or OVVs.

Some of the properties of OVVs are shared by 'BL Lac objects', named after the prototype of the class, BL Lacertae, which was originally identified as a highly variable star, as one can conclude from its variable-star name. BL Lac objects are distinguished by the absence of strong emission or absorption lines in their spectra. Weak stellar absorption features or weak nebular emission lines often can be detected in very high signal-to-noise ratio spectra, however.

It is thought that both OVVs and BL Lacs are those AGNs which have a strong relativistically beamed component close to the line of sight. Collectively, OVVs and BL Lacs are sometimes referred to as 'blazars'.† All known blazars are radio sources.

† The term 'blazar' was introduced by E. Spiegel in his talk following the conference banquet at the Pittsburgh Conference on BL Lac Objects in April 1978 (see Burbidge and Hewitt 1992).

2.6   Narrow-Line X-Ray Galaxies

Some of the sources identified in the early X-ray surveys of the sky were subsequently found to be either Seyfert galaxies or galaxies with some Seyfert-like characteristics. The latter sources were characterized by the same high-excitation emission lines as Seyfert galaxies, but with lower luminosities than typical of Seyferts (Ward *et al.* 1978). These galaxies, which are sometimes called 'narrow-line X-ray galaxies' (NLXGs), are Seyfert galaxies whose optical spectra are heavily reddened and extinguished by dust within the galaxy.

2.7   Related Phenomenology

In this section, we note a few other classifications which one often finds in the AGN literature. The following do not themselves constitute subclasses of AGNs, however.

2.7.1   Starburst Galaxies

A significant fraction of galaxies show the signatures of recent large-scale star formation activity. Such galaxies are known widely as 'starburst galaxies'. These are generally characterized by relatively blue colors and strong H II-region-type emission-line spectra (due to a large number of O and B-type stars) and relatively strong radio emission (due to recent supernova remnants). In some cases, the starburst is apparently confined to an unresolved region at the galactic center, which looks very much like an active nucleus. These 'nuclear starbursts' are typically around 10 times brighter than the giant H II-region complexes seen in normal spirals and are thus distinct from otherwise inactive late-type spirals. The relationship between AGNs and nuclear starbursts is not clear; there is some speculation that there is evolution between the two types of phenomena. A more extreme view is that they are different manifestations of the same phenomenon.

2.7.2   Markarian Galaxies

A large number of the Seyfert galaxies presently known were identified in the low-dispersion spectroscopic survey of Markarian and his colleagues at the Byurakan Observatory in Armenia (see Lipovetsky, Markarian, and Stepanian 1987). The Byurakan survey was done with an objective prism on a 1-m telescope, at a dispersion of $\sim 1800\,\text{Å}\,\text{mm}^{-1}$. These spectra were used to identify UV-excess objects. Huchra (1977) estimates that about 11% of the objects in the Markarian catalogs are Seyfert galaxies, $\sim 2\%$ are Galactic stars (mostly very hot white dwarfs), another $\sim 2\%$ are QSOs and BL Lac objects, and the rest are galaxies which are rather blue for their morphological type (blue compact dwarf galaxies and starburst galaxies).

2.7.3 Zwicky Galaxies and N Galaxies

Luminous Seyfert galaxies have turned up as members of two types of morphologically selected samples of galaxies. Zwicky identified blue compact galaxies as a distinct morphological type, and many of these turned out to be high-luminosity Seyfert galaxies. On the basis of spectroscopic and morphological considerations, these are 'almost QSOs'. For example, the compact blue galaxies II Zw 136 = 2130+099 (the 136th entry in Zwicky's second catalog) and I Zw 1 = 0051+124 have absolute B magnitudes $M_B \approx -21.53 + 5 \log h_0$ and $M_B \approx -21.80 + 5 \log h_0$, respectively, which by the modern definition fall into the QSO class.

Morgan (1958) identified galaxies which seemed to have such strong nuclei that they appear to be nearly stellar on photographs, although it is clear on close inspection that these are not point sources. These were designated as 'N' (for 'nuclear') galaxies. Several of these have turned out to be Seyferts or BLRGs, the latter in particular as these show no structure in their stellar components.

2.7.4 Ultraluminous Far-Infrared Galaxies

Observations with the *Infrared Astronomical Satellite* (*IRAS*) led to the identification of many highly luminous galaxies that emit a substantial fraction of their energy in the far infrared, i.e., at $\lambda > 10 \, \mu$m (Soifer, Houck, and Neugebauer 1987), although the existence of such luminous far-IR sources had been known previously (e.g., Rieke and Low 1972). Ultraluminous far-infrared galaxies have $L(8-1000 \, \mu\text{m}) \gtrsim 10^{12} \, L_\odot$ and have far-infrared luminosities that exceed their optical luminosities by a factor of 10 or more. Many extragalactic *IRAS* sources are starburst galaxies. Some *IRAS* sources are identified with known AGNs, and in other cases the *IRAS* observations led to identification of AGNs not previously known. The far-infrared emission in these sources is thermal radiation from dust (at $T \approx 100$ K or less) that is heated either by massive star formation or by a 'hidden' AGN which we do not observe directly through the dust.

2.8 The Relationship Between Seyferts and Quasars

By virtually any criterion, the properties of quasars and Seyfert galaxies show considerable overlap. The highest-luminosity Seyfert galaxies are indistinguishable from quasars. Even though early investigators understood that quasars and Seyferts had some similar characteristics, it was not until the mid-1970s that most astronomers began to accept the idea that the Seyferts and quasars formed a continuous sequence in luminosity, and in fact were probably the same phenomenon. A fascinating discussion of this issue is given by Weedman (1976). Weedman notes that part of the problem is that the first Seyferts and the first quasars to be discovered were in both cases extremes of the class, which magnified the apparent differences between them. The first known

Seyferts were those identified in Seyfert's original paper, which were nearby bright galaxies, with only slightly unusual cores. The first quasars discovered, on the other hand, were the 3C objects, which are unusual in that they are radio loud, and many of them are OVVs. The arguments that there are fundamental differences between quasars and Seyferts were based on the following points (see Burbidge and Burbidge 1967):

- *Variability.* Although Seyferts were discovered first, it was not until several years later that Fitch, Pacholczyk, and Weymann (1967) published the first report of nuclear variability in a Seyfert galaxy. Prior to this, variability had not been detected because it had not been searched for, since there was no reason to believe it should occur. Variability was, however, known to be common among quasars.
- *Luminosity.* The first known quasars tended to be very luminous, whereas the first Seyferts known were intrinsically faint. The earliest samples studied showed no overlap in luminosity.
- *Emission-line strengths.* Seyferts seemed to have strong lines relative to the continuum (i.e., larger equivalent widths), and quasar lines seemed weaker (i.e., smaller equivalent widths). This argument, however, was based on the comparison of the low-redshift lines in Seyferts (Balmer lines and strong forbidden lines) with the UV lines in high-redshift quasars. The argument that the emission-line equivalent widths in QSOs and Seyferts are different is not so obviously true if one compares the same lines in each object (see, however, Chapter 5). For AGNs *in general*, the optical lines have larger equivalent widths, as is seen very clearly in Table 1.1, simply because there is relatively less continuum emission per unit wavelength under the optical lines.

Simple arguments (see Chapter 3) indicate that the energy density in QSOs is absurdly high, particularly in the case of blazars (§4.5), which we now believe are relativistically beamed sources and thus have lower intrinsic luminosities than we would infer by assuming that their emission is isotropic. The energy arguments that arose primarily on the basis of observations of the beamed sources (radio-loud QSOs and blazars) provided momentum for a school of thought which held that quasars were in fact relatively local objects, perhaps ejected by our own Galaxy, and that the observed redshifts were not due to the Hubble expansion, but had some non-cosmological origin. The contention that 'new physics' was required to explain QSOs still appears in some of the published literature although the original basis for these arguments has been discredited. While it remains true that we do not understand the energy source in quasars, we do not have to explain away the very existence of such high-luminosity sources, as they are plainly seen in nearby Seyferts where we can detect the surrounding stars. There is no question about the distance to Seyferts (beyond the uncertainty in the Hubble constant), and the Seyferts clearly contain these high-luminosity nuclei. Weedman's premise is that the entire history of quasar astronomy might have been different if the Zwicky or Markarian galaxies had been observed before the 3C sources, as indeed nearly happened; the extension of the Seyfert phenomenon to higher luminosities would have then been natural, and it is only an additional small step to reach the quasars once one has come to grips with the high-luminosity Seyferts.

The fact that Seyferts and quasars seem to form a continuous sequence in luminosity should not be construed to mean that there are *no* physical differences between them. We may indeed find that real physical differences appear, depending on the luminosity of the central source.

3 The Black-Hole Paradigm

The fundamental question about AGNs is how the energy that is detected as radiation is generated. Essentially the problem is that an AGN produces as much light as up to several trillion stars in a volume that is significantly smaller than a cubic parsec. The current paradigm, or working model, for the AGN phenomenon is a 'central engine' that consists of a hot accretion disk surrounding a supermassive black hole. Energy is generated by gravitational infall of material which is heated to high temperatures in a dissipative accretion disk. The physical arguments which underlie this view are very basic, and date back at least as far as Zel'dovich and Novikov (1964) and Salpeter (1964). In this chapter, we will review the fundamental arguments that have led to this picture.

3.1 Mass of the Central Object

The mass of the central source can be estimated in a simple fashion, by assuming only isotropy and that the source is stable. For simplicity, we consider the case of a completely ionized hydrogen gas. In order to avoid disintegration, the outward force of radiation pressure must be counterbalanced by the inward force of gravity. The outward energy flux at some distance r from the center is $F = L/4\pi r^2$, where L is the luminosity (ergs s^{-1}) of the source. Noting that the momentum carried by a photon (energy $E = h\nu$) is E/c, the outward *momentum* flux, or pressure, is thus

$$P_{\text{rad}} = \frac{F}{c} = \frac{L}{4\pi r^2 c} \tag{3.1}$$

and the outward radiation force on a single electron is thus obtained by multiplying by the cross-section for interaction with a photon:

$$\mathbf{F}_{\text{rad}} = \sigma_c \frac{L}{4\pi r^2 c} \hat{\mathbf{r}}, \tag{3.2}$$

32

where σ_e is the Thomson scattering cross-section† and $\hat{\mathbf{r}}$ is a dimensionless unit vector in the radial direction. The gravitational force acting on an electron–proton pair (masses m_e and m_p, respectively) by a central mass M is of course $\mathbf{F}_{\text{grav}} = -GM(m_p + m_e)\hat{\mathbf{r}}/r^2 \approx -GMm_p\hat{\mathbf{r}}/r^2$. The inward gravitational force acting on the gas must balance or exceed the outward radiation force if the source is to remain intact, so it is required that

$$
\begin{aligned}
|\mathbf{F}_{\text{rad}}| &\leq |\mathbf{F}_{\text{grav}}| \\
\frac{\sigma_e L}{4\pi c r^2} &\leq \frac{GMm_p}{r^2} \\
L &\leq \frac{4\pi G c m_p}{\sigma_e} M \\
&\approx 6.31 \times 10^4 \, M \text{ ergs s}^{-1} \\
&\approx 1.26 \times 10^{38} \, (M/M_\odot) \text{ ergs s}^{-1}.
\end{aligned}
\tag{3.3}
$$

Equation (3.3) is known as the Eddington limit, and can be used to establish a minimum mass, the Eddington mass M_E, for a source of luminosity L. In units appropriate for AGNs, we can write this as

$$
M_E = 8 \times 10^5 L_{44} \, M_\odot,
\tag{3.4}
$$

where L_{44} is the central source luminosity‡ in units of 10^{44} ergs s^{-1}, which is characteristic of a luminous Seyfert galaxy. For a typical quasar luminosity $L_{\text{QSO}} \approx 10^{46}$ ergs s^{-1}, a central mass in excess of $\sim 10^8 \, M_\odot$ is required. Equivalently, we define the Eddington luminosity

$$
L_E = \frac{4\pi G c m_p}{\sigma_e} M,
\tag{3.5}
$$

which can be thought of as the maximum luminosity of a source of mass M that is powered by spherical accretion.

It is possible to estimate the mass of the central source more directly through an application of the virial theorem (eq. 1.1). Reverberation mapping of the broad emission-line region in AGNs (§5.5) can provide information on both the spatial extent and velocity field of the line-emitting gas. By assuming only that the lines are Doppler-broadened by motions in the gravitational field, a mass estimate of the central source can be obtained. The values obtained are typically somewhat more than an order of magnitude larger than M_E.

† The Thomson cross-section is

$$
\sigma_e = \frac{8\pi}{3} \left(\frac{e^2}{m_e c^2} \right)^2 = 6.65 \times 10^{-25} \text{ cm}^2.
$$

While radiation pressure also acts on the protons, the force on the protons is lower by a factor of $(m_p/m_e)^2 \approx 3 \times 10^6$ because of their higher inertia. The protons and electrons are electromagnetically coupled so charge separation does not occur.

‡ Note that this is a *bolometric* luminosity, since the electron-scattering cross-section is frequency independent below the Klein–Nishina limit (~ 0.5 MeV).

3.2   Fueling Quasars

As in stars, the fundamental process at work in an active nucleus is the conversion of mass to energy. This is done with some efficiency η, so the energy available from a mass M is $E = \eta M c^2$. The rate at which energy is emitted by the nucleus ($L = dE/dt$) gives us the rate at which energy must be supplied to the nuclear source by accretion,

$$L = \eta \dot{M} c^2, \tag{3.6}$$

where $\dot{M} = dM/dt$ is the mass accretion rate. Thus, to power a typical AGN requires an accretion rate

$$\dot{M} = \frac{L}{\eta c^2} \approx 1.8 \times 10^{-3} \left(\frac{L_{44}}{\eta} \right) M_\odot \ \text{yr}^{-1}. \tag{3.7}$$

The viability of accretion as a source of energy (i.e., conversion of gravitational potential energy to radiation) relative to other mechanisms (such as nuclear processes) depends on how large the efficiency factor η can be made. The potential energy of a mass m a distance r from the central source of mass M is $U = GMm/r$. The rate at which the potential energy of infalling material can be converted to radiation is given by

$$L \approx \frac{dU}{dt} = \frac{GM}{r} \frac{dm}{dt} = \frac{GM\dot{M}}{r}, \tag{3.8}$$

where $\dot{M}$ is the mass accretion rate (i.e., mass crossing radius r per unit time). Equations (3.6) and (3.8) show that $\eta \propto M/r$, which is a measure of the 'compactness' of a system. The compactness is maximized in the case of a black hole, whose size we can define in terms of the Schwarzschild radius R_S, which is the event horizon for a non-rotating black hole. This provides a characteristic scale for the object

$$
\begin{aligned}
R_S \ &= \ \frac{2GM}{c^2} \\
&\approx \ 3 \times 10^{13} \ M_8 \ \text{cm} \\
&\approx \ 10^{-2} \ M_8 \ \text{light days},
\end{aligned}
\tag{3.9}
$$

where M_8 is the black hole mass in units of $10^8 \ M_\odot$.

If we ignore relativistic effects, the energy available from a particle of mass m falling to within, say, $5R_S$ (which is about where most of the optical/UV continuum radiation is expected to originate (§3.3)†), is

$$U = \frac{GMm}{5R_S} = \frac{GMm}{10GM/c^2} = 0.1mc^2. \tag{3.10}$$

This oversimplified calculation suggests that $\eta \approx 0.1$, which is an order of magnitude more efficient than fusion of hydrogen to helium, for which $\eta = 0.007$, and is within a factor of ten of the annihilation energy mc^2. For $\eta = 0.1$, eq. (3.7) shows that the accretion rate for even fairly high-luminosity sources, say, $L_{QSO} \approx 10^{46}$ ergs s^{-1}, is only

---

† The innermost stable orbit around a non-rotating (Schwarzschild) black hole is at $3R_S$.

$\dot{M} \approx 2\, M_\odot\, \mathrm{yr}^{-1}$. The value of the efficiency factor η is very uncertain, and depends on the details of how the accretion actually occurs. However, $\eta \approx 0.1$ is of the right order of magnitude and suitable for the calculations here.

We identify the Eddington accretion rate $\dot{M}_E$, which is the mass accretion rate necessary to sustain the Eddington luminosity L_E (eq. 3.5),

$$
\begin{aligned}
\dot{M}_E &= \frac{L_E}{\eta c^2} \\
&\approx 1.4 \times 10^{18} (M/M_\odot)\, \mathrm{g\,s}^{-1}. \\
&\approx 2.2 M_8\, M_\odot\, \mathrm{yr}^{-1}.
\end{aligned}
\tag{3.11}
$$

In the simple spherical accretion model described here, $\dot{M}_E$ represents a maximum possible accretion rate for the mass M. This critical rate can easily be exceeded, however, with models that are not spherically symmetric. For example, the Eddington rate can be exceeded if the mass accretion occurs primarily equatorially in a disk, but the radiation emerges primarily along the disk axis.

The major problem with fueling a quasar by gravitational accretion is thus not the energy requirement, but angular momentum considerations, since the accretion disk is so small. Certainly, the fact that both the radio and extended emission-line structures show apparent axial rather than spherical symmetry suggests that rotation is important at some level. Infalling gas must lose most of its angular momentum before reaching the accretion disk, where further angular momentum transfer can occur through viscosity. To illustrate this, consider a particle in a circular orbit in the solar circle around the Galactic center. The angular momentum per unit mass is $|\mathbf{L}|/m = (GMr)^{1/2}$, where M is the mass interior to r, i.e., $M = 10^{11}\, M_\odot$ and $r = 10\,\mathrm{kpc}$. If this particle is moved to within $\sim 0.01\,\mathrm{pc}$ of a $10^7\, M_\odot$ central black hole, where viscosity might become important (i.e., in the vicinity of the emission-line regions), its angular momentum per unit mass must decrease to $(10^7 \times 0.01\,\mathrm{pc}/10^{11} \times 10^4\,\mathrm{pc})^{1/2} \approx 10^{-5}$ of its initial value. It is for this reason that gravitational interactions with other galaxies (Chapter 8) are sometimes suspected of playing a major role in fueling AGNs, since this can provide a mechanism for removing angular momentum from the gas in the host galaxy.

It may be that the gas that fuels the accretion disk is provided by stars. In order to make use of gas supplied in this way, the star needs to be tidally disrupted rather than being 'swallowed whole' by the black hole, and this turns out to place an upper limit on the mass of the black hole. A star of mass density ρ_* near a massive body of density ρ_{BH} and radius R can approach no closer than the familiar Roche limit

$$
r_R = 2.4 \left(\frac{\rho_{BH}}{\rho_*} \right)^{1/3} R
\tag{3.12}
$$

without being tidally disrupted. Thus, to ensure that a star is tidally disrupted before it crosses the event horizon requires that $r_R > R_S$. By writing the mass density of the black hole in terms of the total mass M within the Schwarzschild radius R_S, eq. (3.12) can be written

$$\frac{r_R}{R_S} = 2.4 \left(\frac{3M}{4\pi R_S^3 \rho_*} \right)^{1/3} > 1, \tag{3.13}$$

which with eq. (3.9) becomes

$$M < 0.64 \left(\frac{c^6}{G^3 \rho_*} \right)^{1/2} \approx 5 \times 10^8 \rho_*^{-1/2} \, M_\odot. \tag{3.14}$$

Thus, objects of stellar density ($\rho_* \approx 1 \, \mathrm{g\,cm}^{-3}$) are tidally disrupted outside the Schwarzschild radius by black holes with $M \lesssim 10^8 \, M_\odot$. For more massive black holes, the Roche limit is smaller than R_S and stars can cross the event horizon without being tidally destroyed. In this case, the mass of the black hole grows through accretion and *gravitational* radiation is produced, but no significant *electromagnetic* radiation results.

Presumably, the gas lost from tidally disrupted stars dissipates energy through shocks and radiation, but conserves angular momentum, which leads to the formation of an accretion disk.

3.3   Accretion-Disk Structure

The detailed structure of the accretion disk depends on a variety of parameters, such as the magnetic field strength and the accretion rate, and the presence or absence of a disk corona or jets. Moreover, the nature of the viscosity is not understood, and the role of thermal instabilities is unclear. Discussions of the basic structure of AGN accretion disks are provided by Blandford (1985) and Begelman (1985) and thorough summaries of accretion theory are given by Frank, King, and Raine (1992) and by Treves, Maraschi, and Abramowicz (1988). Here we consider only a few of the basic properties of thin accretion disks, in particular those that lead to a simple prediction of the emitted spectrum and are not dependent on the unknown viscosity.

It is assumed that indeed the luminous energy of the AGNs is derived by accretion, and it is further assumed that the energy of a particle at distance r from the central source is dissipated locally and that the medium is optically thick. In this case we can approximate the local emission as blackbody. Gravitational potential energy is released at the rate $GM\dot{M}/r$; from the virial theorem, half of this goes into heating the gas, and the other half is radiated away at rate L. Thus,

$$L = \frac{GM\dot{M}}{2r} = 2\pi r^2 \sigma T^4, \tag{3.15}$$

where the σT^4 is as usual the energy radiated per unit area, the factor πr^2 is the area of the disk, and the preceding factor of two accounts for the fact that the disk has two sides. Rearranging this to solve for the temperature at r, we obtain

$$T = \left(\frac{GM\dot{M}}{4\pi\sigma r^3} \right)^{1/4}. \tag{3.16}$$

A more correct derivation takes into account how the energy is dissipated in the disk through viscosity, which is a consequence of work being done by viscous torques. This yields

$$T(r) = \left[\frac{3GM\dot{M}}{8\pi\sigma r^3} \left\{ 1 - \left(\frac{R_{in}}{r} \right)^{1/2} \right\} \right]^{1/4}, \tag{3.17}$$

where R_{in} defines the inner edge of the disk. For $r \gg R_{in}$, this can be simplified and put in terms of R_S (which is certainly a lower limit to R_{in}),

$$T(r) \approx \left[\frac{3GM\dot{M}}{8\pi\sigma R_S^3} \right]^{1/4} \left(\frac{r}{R_S} \right)^{-3/4}. \tag{3.18}$$

By using eq. (3.9), we can write this as

$$T(r) \approx \left[\frac{3c^6}{64\pi\sigma G^2} \right]^{1/4} \dot{M}^{1/4} M^{-1/2} \left(\frac{r}{R_S} \right)^{-3/4}. \tag{3.19}$$

This can also be written in terms the Eddington accretion rate (eq. 3.11) for masses appropriate for AGNs as

$$T(r) \approx 6.3 \times 10^5 (\dot{M}/\dot{M}_E)^{1/4} M_8^{-1/4} \left(\frac{r}{R_S} \right)^{-3/4} \text{ K.} \tag{3.20}$$

For a disk surrounding a $10^8 M_\odot$ black hole that is accreting at the Eddington rate, the emission from the inner part of the disk is maximized (by setting the derivative of the Planck function $dB_\nu/d\nu = 0$) at a frequency

$$\nu_{max} = \frac{2.8kT}{h} \approx 3.6 \times 10^{16} \text{ Hz,} \tag{3.21}$$

which corresponds to a wavelength of $\sim 100 \text{ Å}$ or to a photon energy of $\sim 100 \text{ eV}$, i.e., in the extreme ultraviolet or soft X-ray region of the spectrum. Thus, the thermal emission from an AGN accretion disk is expected to be relatively prominent in the ultraviolet spectrum. The peak temperature one would expect scales like $M^{-1/4}$, so a naïve prediction is that if AGN are all accreting matter at about the same rate (relative to the Eddington rate), the peak temperature of the accretion disk will be lower for more massive black holes. In the case of stellar-mass ($M \approx 1 M_\odot$) black holes, eqs. (3.20) and (3.21) show that the spectrum will peak at around 10^{18} Hz or so, which is why Galactic black-hole candidates are strong X-ray sources.

The actual value of the accretion rate relative to $\dot{M}_E$ and the opacity of the accreting material determines the basic accretion-disk structure. At low accretion rates, $\dot{M}/\dot{M}_E \lesssim 1$, and high opacities, the accretion disk is thin, i.e., the physical height of the disk is small compared to its diameter, and the disk radiates at high efficiency ($\eta \approx 0.1$). The thin disk structure means that the rate at which heat is advected inward is negligible compared to the rate at which it is radiated away in the vertical direction. Thus, the emitted spectrum is a composite of optically thick thermal emission spectra over the range of temperatures that persist through the disk. The X-ray emission arises primarily from the hottest, innermost part of the disk, and the UV and optical continua are dominated by emission from farther out in the disk.

At high accretion rates, $\dot{M}/\dot{M}_E \gg 1$, the upward flowing radiation is partially trapped by the accreting material, and the disk expands vertically into a 'radiation torus', or thick disk, which radiates inefficiently, approximately as $(\dot{M}/\dot{M}_E)^{-1}$ ($\ll 1$). In this case, energy is advected inward faster than the disk can cool radiatively. The structure of such a disk is very much like that of an early-type star, with the opacity primarily due to electron scattering. Because heat transport in the radial direction is non-negligible, the emitted spectrum is close to a single-temperature blackbody spectrum with a temperature of a few times 10^4 K.

At very low accretion rates ($\dot{M}/\dot{M}_E \lesssim 0.1$), the disk becomes optically thin and it is also possible for a stable, two-temperature structure, an 'ion torus', to develop because of the inability of the inner regions of the disk to cool efficiently if the electrons and ions are thermally decoupled. In this case, the ion temperature can reach a value given by the virial theorem ($2K + U = 0$, where $K = 3kT/2$). We thus define the virial temperature T_{vir} through the equation

$$kT_{vir} = \frac{GMm_p}{3r} = \frac{m_p c^2}{6} \left(\frac{r}{R_S} \right)^{-1} \approx 160 \left(\frac{r}{R_S} \right)^{-1} \text{ MeV}, \qquad (3.22)$$

or $T_{vir} \approx 2 \times 10^{12} (r/R_S)^{-1}$ K. Ion tori are suspected of playing a major role in producing jets as the magnetic field of the central source will be frozen into the ionized torus, creating a rapidly rotating field with an axis parallel to the angular momentum vector of the disk. The strong field can collimate the outflow of charged particles, which could lead naturally to jet structures. It is important to remember that magnetic field lines cannot be anchored to a black hole itself, since field lines that cross the event horizon are pinched off in a light-crossing time. An ion torus may provide a simple mechanism of anchoring a strong magnetic field in the vicinity of the black hole.

The principal problem in relating theoretical accretion disks to observations is that the emergent spectrum depends on the details of the accretion-disk structure, and there are many free parameters, such as magnetic field strength, disk inclination, and the extent to which the radiation is reprocessed by the hot corona which is thought to exist above the disk (see Chapter 4). The models provide enough flexibility that by fine tuning the parameters a wide range of possible observed spectra can be produced.

3.4   Alternatives to Black Holes

While the black-hole model for the AGN central engine has a high degree of plausibility, it remains unproven by any reasonable scientific standard. The wide acceptance of this model is at least in part due to the absence of serious contenders. The early days of AGN research (cf. Burbidge and Burbidge 1967) produced a wide variety of remarkable models that were designed to account for the enormous energy output of high-luminosity QSOs, but most of these models have fallen by the wayside as they failed in one way or another to account for various AGN characteristics or were found to be physically untenable. Within the last few years, however, stronger evidence for the existence of supermassive ($M > 10^7 M_\odot$) objects in the centers of galaxies has emerged as a result of (a) dynamical studies of gas in the nucleus of M87 (Ford *et al.* 1994), (b) megamaser kinematics in M106 (Miyoshi *et al.* 1995), which is now well-established as a *bona fide* AGN (Wilkes *et al.* 1995), (c) the detection of very broad, gravitationally redshifted X-ray emission lines in MCG-6-30-15 (Tanaka *et al.* 1995, §4.2), and (d) reverberation mapping of the broad-line regions in AGNs (e.g., Korista *et al.* 1995, §5.5).

Probably the only alternative model that has not been definitively discredited is the nuclear starburst scenario of Terlevich and collaborators (e.g., Terlevich *et al.* 1992). In this model, the radiated energy is supplied by young stars in a nuclear star cluster, and the UV/optical variability and broad emission lines are attributed to frequent supernovae. There are a number of potentially serious difficulties with the model, such as its apparent inability to account for either rapid X-ray variability or radio-loud objects. Moreover, a star cluster would have to be unusually compact, since imaging studies with *Hubble Space Telescope* (*HST*) show that AGNs remain unresolved at the highest currently attainable spatial resolution, $0\overset{''}{.}05$. In the case of the nearest known AGN, NGC 4395, this translates to an upper limit on the size of the nucleus of 0.7 pc ($\sim 2 \times 10^{18}$ cm; Filippenko, Ho, and Sargent 1993).

4   Continuum Emission

In contrast to what was believed for the first twenty years of AGN studies, the continuum spectra of AGNs are quite complex. As mentioned in Chapter 1, at least to a low-order approximation the SED of AGNs can be described as a power law of the form $F_\nu \propto \nu^{-\alpha}$, where α is generally between zero and unity. This led to the initial suspicions that this continuum is non-thermal in origin. It is certainly tempting to attribute the bulk of an AGN spectrum to synchrotron emission, primarily because of the broad-band energy characteristics of the emission and because of the similarity of the spectra to known synchrotron sources such as supernova remnants and extended radio sources. By the end of the 1970s, the best working model to produce the broad-band continuum was the synchrotron self-Compton (SSC) mechanism. Given a power-law distribution of energies, relativistic electrons in a magnetic field can produce a synchrotron power-law spectrum over many decades of frequency. Moreover, it is possible in principle to produce the higher-energy emission, all the way up to X-rays, via the SSC process; the SSC process becomes important when the synchrotron radiation density becomes sufficiently high that the emitted photons are inverse-Compton scattered off the very electrons which are responsible for the synchrotron radiation. The major difficulty in understanding whether a particular source of radiation is pure synchrotron emission or SSC is that SSC-boosted photons have the same relative energy distribution as the original photons, thus providing no unique indication of the process. The primary discrepancy between the predictions of the SSC models and the observations is that the continuum should be polarized by at least a few percent, and such high levels of polarization are observed only in the blazar class.

Figure 4.1 shows typical SEDs for luminous radio-loud and radio-quiet normal (non-blazar) AGNs. The highest energies shown in this diagram are at X-ray energies ($\sim 10\,\mathrm{keV}$); X-ray and γ-ray photons do not penetrate the Earth's atmosphere, so high-energy observations are carried out with satellite-based detectors, which have carried out observations to energies as high as $\sim 30\,\mathrm{GeV}$ over extended periods of time. AGNs can be observed indirectly at even higher energies from the ground (up to $\sim 10^3\,\mathrm{GeV}$) through Čerenkov emission that results from high-energy photons hitting the Earth's atmosphere.

At energies below 1 keV, there are two significant gaps in the spectral coverage where data are simply not available, one in the extreme ultraviolet (EUV) part of the spectrum ($\sim 10^{16}\,\mathrm{Hz}$) and the other in the millimeter-wavelength regime. The EUV gap is due primarily to the opacity of the interstellar medium in our own Galaxy; absorption by neutral hydrogen in the Galactic disk makes detection of any extragalactic source

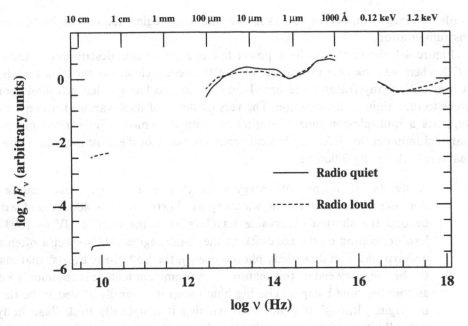

Fig. 4.1. Mean spectral energy distributions (SEDs) for a sample of radio-quiet (solid line) and radio-loud (dashed line) QSOs (from Elvis *et al.* 1994). The flux scale has been arbitrarily normalized at a wavelength of 1 μm.

virtually impossible between 912 Å (the Lyman continuum edge) and ∼ 100 Å. It is not coincidental that this is the part of the spectrum to which the AGN emission lines are most sensitive (Chapters 5 and 6) since both the AGN line-emitting clouds and our own Galactic interstellar medium have a large neutral hydrogen content. This gap therefore leads to considerable uncertainties in our understanding of the physics of AGN line-emitting regions. The long-wavelength gap occurs for several reasons. At the short-wavelength end, the primary factor that limits spectral coverage in the infrared and at longer wavelengths is the opacity of the Earth's atmosphere on account of water vapor absorption at wavelengths between ∼ 1 μm and ∼ 300 μm. The data at wavelengths just longward of ∼ 1 μm are obtained through a limited number of nearly transparent atmospheric 'windows' in the near IR that extend to ∼ 20 μm. All of the data in Fig. 4.1 in the wavelength range 12–100 μm were obtained above the atmosphere with the *Infrared Astronomical Satellite* (*IRAS*). At wavelengths longward of ∼ 300 μm (10^{12} Hz), the opacity of the Earth's atmosphere is low enough to permit ground-based observations to wavelengths as long as ∼ 150 m (2×10^6 Hz), beyond which the Earth's ionosphere precludes transmission of long-wavelength radiation. The absence of data at the long-wavelength end ($\lambda \gtrsim 300$ μm) of the millimeter gap is due primarily to the lack of sensitive detectors in this range. A suitable combination of accurate telescope surface figures and collecting areas large enough for systematic studies of large numbers of AGNs in the millimeter regime requires highly specialized instrumentation.

Figure 4.1 shows that while a power law is a reasonable description of the AGN SED when we look over several decades of frequency, closer examination reveals that it overlooks many features, i.e., broad depressions and bumps, that provide important clues to the origin of the emission. The very existence of these various features strongly suggests a multiple-component continuum, with the emission in various continuum bands dominated by different physical processes. Some of the most important observed features include the following:

- A significant amount of energy is emitted in a strong, broad feature that dominates the spectrum at wavelengths shortward of $\sim 4000\,\text{Å}$ and extends beyond the shortest observable wavelengths in the satellite UV ($\sim 1000\,\text{Å}$ or less, depending on the redshift). In the X-ray region, AGN spectra often show a sharp rise with decreasing photon energy, the 'soft X-ray excess', that may be the high-energy end of this feature. This prominent feature is commonly known as the 'big blue bump'.† The big blue bump is generally agreed to be thermal in origin, although it is not clear whether it is optically thick (blackbody) or optically thin (free–free) emission. Many proponents of the optically thick interpretation ascribe the big blue bump to the purported accretion disk.
- Most AGNs show a local minimum in their SED in the vicinity of $1\,\mu\text{m}$. This is thought to represent the minimum between a hot thermal spectrum (the big blue bump) and a cool thermal spectrum due to emission by warm ($T \lesssim 2000\,\text{K}$) dust grains. The smooth bump at wavelengths longward of $\sim 1\,\mu\text{m}$ is often referred to as 'the IR bump'.
- The SEDs of radio-quiet AGNs decrease rapidly at low energies. The point at which this abrupt decrease begins is known as the 'submillimeter break'.

In this chapter, we discuss briefly some of the basics of continuum emission, in terms of both phenomenology and possible interpretations. Because of the complex nature of the continuum mechanisms, one must be very careful in discussing models since it is easy to over-specify the solution – there are a sufficient number of free parameters in most models that any particular observed spectrum can be matched in detail, although not with a unique solution. The continuum emission alone does not contain unambiguous signatures that immediately lead us to prefer one model over another. For this reason, we will confine the discussion to generalities.

At a fundamental level, the major question is how much of the AGN spectrum is thermal emission, and how much is non-thermal? By thermal emission, we mean radiation due to particles which have attained a Maxwellian velocity distribution by collisions. Non-thermal emission, by contrast, is due to particles whose velocities are

---

† There is a weaker feature superposed on the big blue bump between ~ 2000–$4000\,\text{Å}$, known as the 'small blue bump'. This feature is attributable to a combination of Balmer continuum emission and blends of Fe II emission lines arising in the broad-line region (see Chapter 5).

not described by a Maxwell–Boltzmann distribution, for example, particles which give rise to synchrotron power-law spectra, since the kinetic energy distribution of these particles must also be a power law.

A closely related question is how much of the observed emission is *primary*, i.e., due to particles which are powered directly by the central engine, e.g., synchrotron-emitting relativistic particles, or thermal emission from an accretion disk, and how much is *secondary*, i.e., emitted by gas which receives its energy from the primary processes and re-radiates the emission, e.g., free–free emission from a photoionized (or collisionally ionized) gas.

Part of the reason the answer to these questions is so fundamental has to do with the isotropy of thermal versus non-thermal radiation. Whereas optically thin thermal radiation is isotropic, non-thermal radiation such as synchrotron radiation can be highly directed even in the optically thin case. This is an especially important consideration if a significant amount of the radiation originates in material moving with bulk relativistic velocities, such as in jets – if the radiation is highly directed, (1) we cannot use the observed fluxes and distances to infer the luminosity, and (2) we will not have a true picture of the relative importance of various processes by which the AGN emits radiation. For example, if our line of sight is close to the axis of a relativistic jet, the non-thermal emission from the jet might dominate the observed spectrum since the emission is highly beamed in the direction of the electron motion, but the total amount of energy in the beam may be far less than that emitted by thermal processes, which is radiated isotropically. Indeed, this is thought to be the characteristic that distinguishes blazars from other AGNs. Blazar spectra are sufficiently different from those of normal AGNs that we discuss them separately (§4.5).

4.1 Ultraviolet–Optical Continuum

As noted above, the dominant feature of the UV/optical spectra of AGNs is the big blue bump, which is attributed to some kind of thermal emission in the range $10^{5\pm1}$ K, depending somewhat on whether the emission arises in an optically thin or optically thick region. It is tempting to ascribe the big blue bump to emission from an accretion disk surrounding a supermassive black hole; eq. (3.21) indicates that continuum emission from the accretion disk will be at a maximum at wavelengths longward of ~ 100 Å or so, and thus the UV/optical spectral range is where we should look for evidence of an accretion-disk spectrum.

We consider first the expected characteristics of an accretion-disk spectrum. As a first approximation, we can assume that the accretion disk radiates locally like a blackbody,

$$B_\nu = \frac{2h\nu^3}{c^2} \frac{1}{e^{h\nu/kT(r)} - 1},$$ (4.1)

where $T(r)$ is the disk temperature at radius r from the center, as given approximately by eq. (3.18), which we rewrite for simplicity as

$$T(r) \approx T_* \left(r/R_S \right)^{-3/4}. \tag{4.2}$$

The flux emitted per unit area is $F_\nu = \pi B_\nu$. We assume that the disk is geometrically thin (though optically thick) and has radial symmetry so that a ring of width dr at radius r will have a specific luminosity

$$dL_\nu(r) = 2\pi r \cos i \, \pi B_\nu dr, \tag{4.3}$$

where i is the inclination of the disk to the plane of the sky.† The total specific luminosity of the disk is given by integrating over the full range of radii:

$$L_\nu = \frac{4\pi^2 h\nu^3 \cos i}{c^2} \int_{R_{in}}^{R_{out}} \frac{r \, dr}{e^{h\nu/kT(r)} - 1}, \tag{4.4}$$

where R_{in} and R_{out} are the inner and outer boundaries of the disk, respectively. We can predict approximately the emitted spectrum without specifying R_{in} and R_{out} precisely. In the low-frequency regime, $h\nu \ll kT(R_{out})$, all annuli are radiating in the Rayleigh–Jeans limit (i.e., $B_\nu \propto \nu^2$), so the low-frequency integrated spectrum will have this same form. At high frequencies, thermal emission follows the Wien law (i.e., $B_\nu \propto \nu^3 e^{-h\nu/kT}$), so the spectrum is dominated by the highest-temperature regions because of the exponential cut-off. This means that at $h\nu \gg kT(R_{in})$, the spectrum can be approximated by the spectrum of the hottest regions, i.e., $L_\nu \propto \nu^3 \exp\left[-h\nu/kT(R_{in})\right]$. The spectrum at intermediate frequencies, where most of the energy is emitted, can also be approximated fairly simply. The interesting case is where there is a wide range of temperatures in the disk, since otherwise the emitted spectrum will look fairly close to a single-temperature blackbody spectrum. A wide range of temperatures means that $R_{out} \gg R_{in}$, and we consider the intermediate range in frequency $kT(R_{out})/h \ll \nu \ll kT(R_{in})/h$. We then rewrite eq. (4.4) using the substitution

$$x = \frac{h\nu}{kT(r)} = \frac{h\nu}{kT_*} \left(\frac{r}{R_S} \right)^{3/4} \tag{4.5}$$

and note that at $r = R_{in} \approx R_S$, $x \ll 1$ since $h\nu \ll kT(R_{in})$. Furthermore, as $r \to R_{out}$, $x \to \infty$, so we can approximate eq. (4.4) as

$$\begin{aligned}
L_\nu &= \frac{16\pi^2 R_S^2 h\nu^3 \cos i}{3c^2} \left(\frac{kT_*}{h\nu} \right)^{8/3} \int_0^\infty \frac{x^{5/3} \, dx}{e^x - 1} \\
&= 2.4 \times 10^{-18} R_S^2 \cos i \, T_*^{8/3} \nu^{1/3} \ (\mathrm{ergs \, s^{-1}}) \\
&\propto \nu^{1/3}.
\end{aligned} \tag{4.6}$$

An important prediction of the thin accretion-disk model is that most of the radiation observed at ultraviolet wavelengths arises primarily in different parts of the disk than does the optical radiation. We consider as an example a $10^8 \, M_\odot$ black hole that is

---

† This refers to the energy emitted by *one* side of the disk.

accreting at the Eddington rate (i.e., a luminosity $L \approx 10^{46}$ ergs s^{-1} QSO). Equations (3.21) and (3.18) show that the peak contribution to the 1500 Å continuum arises at $r \approx 50R_S$, i.e., around 1.4×10^{15} cm or 0.6 light days from the black hole. The peak contribution to the 5000 Å continuum, however, arises at $r \approx 240R_S$, i.e., around 7.1×10^{15} cm or 2.8 light days.

Is the shape of the UV/optical continuum in AGNs consistent with the thin accretion-disk prediction, eq. (4.6)? This is a controversial point. After allowing for the emission lines (including the small blue-bump contribution) and contamination of the continuum by starlight from the host galaxy, power-law fits to the UV/optical continuum typically yield $F_\nu \propto \nu^{-0.3}$, i.e., the energy per unit frequency is *decreasing* with frequency rather than increasing like $\nu^{1/3}$. However, acceptable fits can be obtained if it is supposed that there are other contributors to the UV through IR continuum. Specifically, the early work that identified the big blue bump with an accretion-disk spectrum (e.g., Shields 1978, Malkan and Sargent 1982) assumed that underlying the blue-bump component is an IR/X-ray power law (of the form $F_\nu \propto \nu^{-1}$, or thereabouts), which is an assumption that subsequently has fallen out of favor as the consensus view has evolved towards a thermal origin for the IR emission (§4.3). However, Malkan (1989) has shown that the combination of a cool thermal spectrum in the IR plus an accretion-disk spectrum in the UV/optical can successfully model AGN continua over this entire spectral range.

While the assumption of blackbody emission is a useful starting point, it is an oversimplification. Many attempts have been made to refine the predictions of the emergent spectrum of accretion disk. In some cases, for example, model stellar atmospheres (although with inappropriately high surface gravities) have been used to approximate the disk spectrum. More complex models have been developed that more correctly deal with rotational, inclination, and relativistic effects and employ more complete radiative transfer treatments (e.g., Czerny and Elvis 1987, Sun and Malkan 1989, Laor and Netzer 1989, Laor, Netzer, and Piran 1990). Important predictions that result from various model calculations are:

- *Polarization of the continuum.* Relativistic effects should result in significant polarization of the accretion-disk continuum. For high-inclination disks, the polarization can be as high as several percent.
- *Lyman discontinuity.* A Lyman edge (i.e., at 912 Å), in either emission or absorption, depending on the details of the model, is expected in any source in which there is a temperature gradient in the vertical direction, primarily because the opacity (and hence the effective depth of the 'photosphere') changes drastically at the Lyman edge.

Neither of these effects has been demonstrated in a satisfactory fashion. In general, AGN optical continua are polarized at less than the 1% level, except for blazars. Lyman edges have been searched for extensively, with only somewhat ambiguous evidence for absorption in some cases (Reichert *et al.* 1988, Antonucci, Kinney, and Ford 1989, Koratkar, Kinney, and Bohlin 1992, Zheng *et al.* 1995a). However, there are a number of ways to avoid, or at least mask, the Lyman edge problem. For example, for a relativistically rotating disk seen at some inclination $i > 0$, the approaching side of the disk is strongly Doppler-boosted and Doppler-broadened, which can greatly diminish the contrast of the Lyman edge. Even in the case of a face-on disk ($i = 0$) it may be possible to smear out the Lyman edge if the disk has a hot corona of electrons, as expected in the case of a thin disk. A problem with this is that the scattering corona should increase the polarization of the continuum to levels higher than observed, so it is not clear that this is the solution to the problem. Indeed, in a few cases where spectropolarimetry around the Lyman edge has been carried out, the observed polarization increases dramatically towards shorter wavelengths, which is in the opposite sense of the prediction that the polarization should drop at wavelengths shortward of the Lyman edge (Koratkar *et al.* 1995). Complete self-consistent radiative transfer calculations of the disk spectrum will be necessary to resolve the Lyman-edge question.

While the basic physics of the continuum production has not yet been identified, important observational constraints are now becoming available through variability studies, many of which have been undertaken as part of emission-line reverberation-mapping experiments (§5.5). A long-term, well-sampled optical continuum light curve for the bright Seyfert 1 galaxy NGC 5548 is shown in Fig. 4.2.

Studies of variability of the UV/optical continuum reveal that (1) the ultraviolet and optical continua vary in phase, (2) the continuum tends to get 'harder' (the spectral index α decreases) as it gets brighter, i.e., the variations are larger at shorter wavelengths, even after accounting for the greater starlight contribution at longer wavelengths, and (3) smaller-scale (few percent) variations that are seen in the UV on short time scales seem to be smoothed out at longer wavelengths. There is no evidence in particular for any periodic variability that might be associated with orbital motions. The continuum variations seem to be completely irregular.

A few attempts have been made to determine the Fourier power spectra of observed light curves, which can potentially identify special time scales for variability or at the very least provide a measure of how much variability power there is on different time scales. The behavior of the light curve $C(t)$ can be characterized by its Fourier transform

$$\tilde{C}(f) = \int_{-\infty}^{\infty} C(t)\, e^{-i2\pi ft}\, dt, \tag{4.7}$$

where f is the temporal frequency (in units such as Hz or days^{-1}). The 'power spectrum' $P(f)$ is formed by multiplying $\tilde{C}(f)$ by its complex conjugate. It is customary to fit the power spectrum with a power-law form, i.e., $P(f) \propto f^n$, at least over moderate ranges in temporal frequency. The total power is obtained by integrating over all possible temporal frequencies. Integration from some fixed temporal frequency to arbitrarily high temporal frequencies (i.e., very fast variations) requires that $n < -1$ for convergence. Similarly, integration from some fixed frequency to arbitrarily low frequencies requires that $n > -1$ to avoid divergence. Thus, a power spectrum cannot

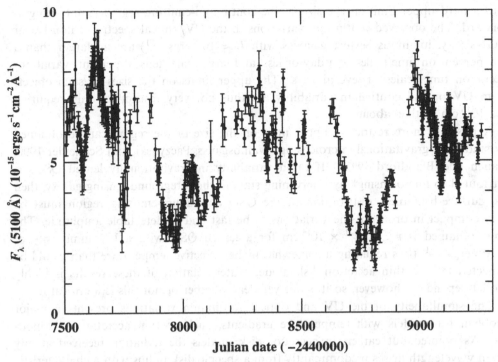

Fig. 4.2. The optical (5100 Å) light curve for the variable Seyfert 1 galaxy NGC 5548 from 1988 to 1993. A constant contribution due to starlight from the host galaxy has been subtracted, so this represents the true nuclear variations. The gaps in the curve occur at times when NGC 5548 is too close to the Sun to observe. Adapted from Peterson (1994).

be characterized by a single index at all frequencies, and the frequencies at which the changes in the power-law index occur can presumably identify fundamental time scales associated with the sources (e.g., at high frequencies, the light-travel time across the source). The UV/optical power spectra of non-blazar AGNs seem to be approximately of the form $P(f) \propto 1/f^{2-3}$, showing that most of the variability is on longer rather than shorter time scales. Low temporal frequencies (corresponding to time scales of years) have not been sufficiently well sampled to determine at what temporal frequency the power spectrum turns over. Our understanding of variations on the shortest time scales (less than days) is also very limited not only because few data exist, but because the high-frequency variations are so small that they are often lost in the observational noise.

In principle, continuum variability affords some limitation on the size of the continuum, based on causality considerations. Simply put, a significant change in the continuum brightness on a short time scale Δt requires that a correspondingly significant fraction of the emitting material be contained within a volume limited by light-travel time considerations to be $r \lesssim c\Delta t$. Translating a time scale for variability into a hard upper limit on the size of the continuum-emitting region is not straight-forward. The observed continuum variations in the UV/optical spectra of non-blazar AGNs (say, luminous Seyfert galaxies with $L \approx 10^{44}$ ergs s^{-1}) are no larger than a few percent on time scales of a day or less, and large-scale (tens of percent) variations occur on time scales of several days. The upper limits on the sizes of such objects from UV/optical continuum variability are thus not very restrictive, only requiring $r \lesssim 10^{16}$ cm or thereabouts.

A possibly more restrictive upper limit on the size of the region can be obtained by studies of gravitational microlensing (Wambsganss, Paczyński, and Schneider 1990, Rauch and Blandford 1991). If rapid variability in gravitationally lensed QSOs is attributed to microlensing[†] by intervening stars in the foreground lensing galaxy, then the surface-brightness distribution of the QSO continuum-emitting region must be very compact in order for the variations to be fast and of detectable amplitude. The limit obtained is $r \lesssim 1.5 h_0^{-1} \times 10^{15}$ cm for a lensed QSO with a UV luminosity of $\sim 10^{46}$ ergs s^{-1}, thus requiring a somewhat higher effective temperature than would be expected from a thin accretion-disk model. Interpretation of these results is highly model dependent, however, so it is not yet clear whether or not this is a critical test.

The simultaneity of the UV and optical continuum variations presents a major problem for models with temperature gradients, such as thin accretion-disk models. As pointed out earlier, in accretion-disk models the radiation received at any given wavelength arises predominantly from a specific disk radius with a characteristic temperature. Thus, if different parts of the disk undergo changes in local emitting conditions (for example, because of disk hydrodynamical instabilities or changes in the accretion rate), then the various continuum bands are not expected to vary simultaneously. The near simultaneity of the UV and optical continuum variations in well-observed Seyfert galaxies requires that the different parts of the disk that provide most of the energy at these respective wavelengths are causally coupled on a time scale so short that information must propagate through the disk at speeds typically greater than $0.1c$ rather than at the sound speed, as expected in a viscous disk.

Given the various difficulties with interpreting the UV/optical spectrum in terms of a thin accretion disk, are there any viable alternatives? One possible explanation is that the UV/optical continuum represents optically thin (i.e., free–free emission or bremsstrahlung) rather than optically thick (blackbody, or modified blackbody) emission, as has been discussed by Barvainis (1993), since either form gives acceptable fits to the observed SEDs (Malkan and Sargent 1982). The primary liability of free–free models for AGN emission is that the efficiency is fairly low, which requires either large emission regions or quite high temperatures ($\sim 10^6$ K). The advantages of this interpretation are several, including (a) in the UV/optical region, the spectral index for free–free emission is consistent with the observed values ($\alpha \approx 0.3$), (b) the polarization

---

[†] An important means of discriminating microlensing from intrinsic variability is that a microlensing event will have the same amplitude at all wavelengths. In general, intrinsic variations in AGNs are larger at higher frequencies.

is expected to be low, and (c) the Lyman edge is expected to be weak as long as the temperature is high ($\sim 10^6$ K). The size constraints do not at this point appear to be in conflict with the UV/optical variability, but rapid X-ray variability, as discussed in the next section, may prove to be more of a problem for the free–free interpretation.

4.2 High-Energy Spectra

High-energy (X-ray and γ-ray) observations are of great importance to an understanding of the AGN phenomenon. Not only do X-rays account for typically $\sim 10\%$ of the bolometric luminosity of AGNs, but their rapid variability indicates that the X-rays provide a probe of the innermost regions of AGNs.

When dealing with high-energy observations, one must bear in mind that in general the spectral resolution is very poor ($R = \lambda/\Delta\lambda \lesssim 10$); the highest spectral resolution obtained to date in high-energy observations of many AGNs is with the *Advanced Satellite for Cosmology and Astrophysics* (*ASCA*) X-ray satellite ($R \lesssim 50$). As a result, features such as absorption or emission lines do not stand out clearly in high-energy spectra, and they are detected by fitting a spectral model that is convolved with the wavelength-dependent instrumental response to the observed data, rather than by direct identification in the observed spectra, which is common at UV, optical, and IR wavelengths. Also, the terminology used in describing the various high-energy regimes (e.g., 0.1–2 keV are 'soft X-rays', 2–100 keV are 'hard X-rays', and energies above 100 keV are 'γ-rays') are artificial distinctions which are imposed by the technology used in the observations. There are other terms that are rather specific to X-ray astronomy as well. For example, whether a gas is 'hot' or 'cold' depends on its temperature relative to the energy of the photons in the band. For a photon energy of 1 keV, the equivalent temperature is $T = h\nu/k \approx 10^7$ K, so a 10^4 K thermal plasma is 'cold' by comparison. Another common difference in X-ray astronomy is that power-law SEDs are usually fitted in units like photons per keV rather than energy per unit frequency. X-ray power-law fits are generally of the form

$$P_E(\text{photons s}^{-1}\,\text{keV}^{-1}) \propto E^{-\Gamma} \propto \nu^{-\Gamma}, \tag{4.8}$$

since photons per second is close to the measured quantity, which is counts per second. In units of energy flux (or luminosity) as we have used so far,

$$
\begin{aligned}
F_\nu &\propto P_E(\text{photons s}^{-1}\,\text{keV}^{-1}) \times h\nu(\text{ergs/photon}) \\
&\propto \nu^{-\Gamma+1} \propto \nu^{-\alpha}.
\end{aligned} \tag{4.9}
$$

To distinguish clearly between these two spectral indices, α is usually referred to as the 'energy index' and Γ ($= \alpha + 1$) is called the 'photon index'. Early X-ray observations (e.g. Mushotzky *et al.* 1980) revealed that over the range 2–20 keV the data can be fitted with a canonical power law of energy index $\alpha \approx 0.7 \pm 0.2$, i.e., with the amount of energy per decade (νF_ν) increasing with frequency. However, more recent, higher-resolution work has revealed the presence of features in the higher-energy part of this band that made the spectrum seem flatter than it actually is. Nandra and Pounds (1994) have shown that a single power law is not a good description of a typical AGN spectrum over this entire range – at higher energies, AGN continua typically rise above an extrapolation of the lower-energy power-law spectrum, as described below. Thus, the power-law component at lower energies is thus somewhat steeper than previously reported, i.e., with $\alpha \approx 0.9$–1.

Observations of faint sources at γ-ray energies have become routinely possible only with the advent of the *Compton Gamma-Ray Observatory* (*CGRO*). The observed γ-ray fluxes of AGNs in the 50–150 keV range fall below the flux expected by extrapolation of the X-ray power law. The γ-ray spectral indices are typically softer than observed in the X-ray region, typically with $\alpha \approx 1.2 \pm 0.2$ (Johnson *et al.* 1994). Fits to the X-ray and γ-ray SEDs of Seyfert galaxies suggest a high-energy cut-off at energies of around a few hundred keV and an SED shape that is consistent with that of the observed γ-ray background (Zdziarski *et al.* 1995), which is probably produced by the integrated emission from these sources. At energies higher than a few hundred keV, with a few notable exceptions such as NGC 4151 (e.g., Maisack *et al.* 1993), only blazar-type (§4.5) objects, i.e., those with a strong beamed component, have been detected.

The relative amounts of energy in the UV and X-rays is sometimes expressed in terms of the (energy) index of a power law that joins the two bands; this should be understood to be nothing more than a simple parameterization of the relative UV/X-ray fluxes, and not a true fit to the spectrum. It is conventional (Tananbaum *et al.* 1979) to define an 'optical/X-ray' spectral index α_{ox} between 2500 Å and 2 keV through eq. (4.9),

$$
\begin{aligned}
\alpha_{ox} &= -\log\left[\frac{F_\nu(2\,\mathrm{keV})}{F_\nu(2500\,\text{Å})}\right] \bigg/ \log\left[\frac{\nu(2\,\mathrm{keV})}{\nu(2500\,\text{Å})}\right] \\
&= -0.384\log\left[\frac{F_\nu(2\,\mathrm{keV})}{F_\nu(2500\,\text{Å})}\right].
\end{aligned}
\tag{4.10}
$$

Since AGNs are detected at a wide variety of redshifts, care must be taken to determine the 2 keV and 2500 Å fluxes in the *rest frame* of the source. It is therefore necessary to correct the observed fluxes at each of these energies to the values that would be measured in the rest frame of the AGN.† Typically, it is found that $\alpha_{ox} \approx 1.4$. Equation (4.10) shows that small differences in α_{ox} translate to large differences in the X-ray/UV flux ratio.

The origin of X-ray and γ-ray emission in AGNs is not understood. In many models, the X-ray emission is produced by inverse-Compton scattering of lower-energy photons by more energetic electrons, although the specific details vary greatly from model to model. The basic idea in at least some of the models is that the UV/optical continuum from the accretion disk is up-scattered (in energy) by inverse-Compton scattering off hot (possibly relativistic) electrons in a corona surrounding the disk, a process commonly referred to in the literature as 'Comptonization' of the input

† This is called the '*K*-correction', and it is discussed in detail in Chapter 10.

(UV/optical) spectrum. In some models, a significant amount of the accretion energy is deposited directly into this hot corona (Haardt and Maraschi 1993), but how the accretion disk and the corona are coupled is not well understood. A purely thermal origin for the X-ray emission, however, cannot be ruled out.

Whereas most of the X-ray emission can be identified as arising in the spatially unresolved nucleus, several Seyfert galaxies have extended structure when observed at soft X-ray energies (e.g., Elvis *et al.* 1990, Wilson *et al.* 1992). The origin of the extended X-ray emission is not clear. Plausible origins include electron-scattered nuclear light and thermal bremsstrahlung from hot gas.

Compton-scattering processes must be at work in active nuclei, simply based on the source sizes and luminosities. Electron–positron pair production (i.e., $\gamma + \gamma \rightarrow e^+ + e^-$) can occur for photon energies above the threshold energy $m_e c^2 = 511$ keV†. The optical depth for photon–photon interactions in a source of size R is $\tau_{\gamma\gamma} = n_\gamma \sigma R$, where n_γ is the number density of photons (cm^{-3}) and σ is the cross-section (cm^2) for pair production. The cross-section for pair production reaches a maximum value close to the Thomson cross-section σ_e just above the energetic threshold for the reaction. The number density of photons with energies near the threshold can be estimated from the energy density (ergs cm^{-3}) of these photons, $U_{\rm rad} \approx L_\gamma / 4\pi R^2 c$, divided by the energy per photon ($\sim m_e c^2$), where L_γ is the γ-ray luminosity of the source. The source will be optically thick to pair-production reactions provided that

$$\tau_{\gamma\gamma} = \left(\frac{L_\gamma}{4\pi R^2 m_e c^3} \right) \sigma_e R \gtrsim 1. \tag{4.11}$$

Rearranging this slightly, we introduce the dimensionless 'compactness parameter' ℓ (Guilbert, Fabian, and Rees 1983) and the condition for pair production to be important becomes

$$\ell \equiv \frac{L_\gamma}{R} \frac{\sigma_e}{m_e c^3} \gtrsim 4\pi. \tag{4.12}$$

By expressing the luminosity in units of the Eddington luminosity (eq. 3.5) and the source size in units of the Schwarzschild radius (eq. 3.9), we can write the compactness parameter as

$$\ell = \frac{L_\gamma}{L_{\rm E}} \left(\frac{4\pi G c m_p M}{\sigma_e} \right) \frac{R_{\rm S}}{R} \left(\frac{c^2}{2GM} \right) \frac{\sigma_e}{m_e c^3}$$

$$= 2\pi \left(\frac{L_\gamma}{L_{\rm E}} \right) \left(\frac{R_{\rm S}}{R} \right) \left(\frac{m_p}{m_e} \right). \tag{4.13}$$

By combining eq. (4.12) and eq. (4.13), we obtain the condition

$$\frac{L_\gamma}{L_{\rm E}} \gtrsim \frac{2m_e}{m_p} \left(\frac{R}{R_{\rm S}} \right) \approx 1.1 \times 10^{-3} \left(\frac{R}{R_{\rm S}} \right). \tag{4.14}$$

---

† The threshold for electron–positron pair production depends on the product of the two photon energies, which must exceed $(m_e c^2)^2$ (Svensson 1990).

Since the size of the X-ray emitting region is of order $10R_S$, this condition can be met in AGNs (Done and Fabian 1989), and production of electron–positron pairs might be expected. The particular relevance of this is twofold. First, the observed power-law slope ($\alpha \approx 0.9$) is consistent with the predictions of saturated† pair-production models without a great deal of fine-tuning (Svensson 1987, Zdziarski *et al.* 1990), which thus lends some credibility to the general model. Second, saturated pair models predict that AGNs will be weaker γ-ray sources than expected from extrapolation of the X-ray power law, as is consistent with the observations.

In addition to the basic power-law form of the X-ray spectra of AGNs, other independent components have been identified from fits to the data. At low energies (primarily at energies $h\nu \lesssim 2\,\mathrm{keV}$), absorption from heavy elements is often detected in AGN spectra (Reichert *et al.* 1985). These primarily are K and L-shell‡ ionization edges of Fe, Mg, Si, and S. By assuming solar abundances, the equivalent hydrogen column densities necessary to produce the observed absorption are of order $\sim 10^{22}\,\mathrm{cm}^{-2}$. The strength of the absorption can vary with time, leading to to the deduction that these are 'warm absorbers', i.e., they have ionization fronts that are variable in position within the absorber and depend on the incident continuum flux (Halpern 1984). Many lower-luminosity AGNs (Seyfert galaxies) also show absorption features in their UV/optical spectra (Ulrich 1988). There is evidence that the UV and X-ray absorption arises in the same clouds (Mathur, Elvis, and Wilkes 1995), although these clouds seem to be too highly ionized to identify with, for example, the standard broad-line emitting clouds (Chapter 5). The UV absorption features in Seyfert galaxies always appear on the short-wavelength (relatively blueshifted) side of the broad emission lines, indicating the absorbing gas is flowing outward from the nucleus. This gas may have some relationship to the broad absorption line (BAL) features seen in some quasars (Chapter 12).

In the soft X-ray region of many AGNs, a steep rise in the spectrum is seen towards low energies (Arnaud *et al.* 1985, Turner and Pounds 1989). This is usually referred to as the 'soft excess'. While its origin is unknown, it has been argued that this feature is the Comptonized Wien tail of the big blue bump (Fig. 4.3). This would require a big blue bump temperature of $\sim 2 \times 10^5\,\mathrm{K}$, which is very high for an accretion disk (cf. eq. 3.20). In some cases, the soft excess may be ascribable at least in part to blended soft X-ray line emission (Turner *et al.* 1991).

At higher energies ($h\nu \gtrsim 10\,\mathrm{keV}$), AGN X-ray spectra also rise above the power-law spectrum, as noted earlier; this 'hard tail' is usually attributed to Compton reflection off a cold gas (i.e., inelastic scattering of high-energy photons off low-energy electrons), perhaps the accretion disk itself (Guilbert and Rees 1988, Lightman and White 1988). The geometry of such a model would require the original or primary X-rays to illuminate the accretion disk from above.

In addition to these broad continuum features, at least one emission line, Fe Kα,

---

† The term 'saturated' means that the γ-ray optical depth is sufficiently high that virtually all γ-rays are converted to electron–positron pairs.

‡ K-shell ionization refers to the process of removing one of the electrons from the innermost ($n = 1$) closed electron shell in the atom. L-shell ionization refers to the $n = 2$ shell.

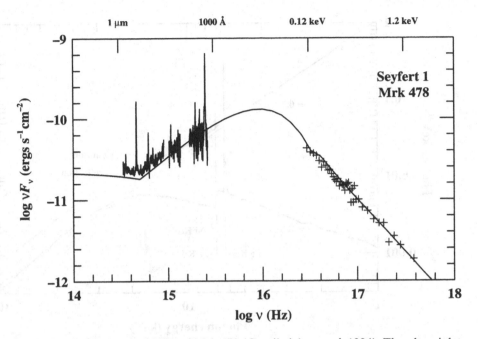

Fig. 4.3. The high-energy SED of Mrk 478 (Gondhalekar *et al.* 1994). The ultraviolet data are from the *International Ultraviolet Explorer* (*IUE*) and the soft X-ray data are from the *Röntgen Astronomical Satellite* (*ROSAT*). The solid line shows a model SED that reproduces the emission-line strengths. These data and the model SED suggest that the 'big blue bump' peaks around 10^{16} Hz (~ 300 Å) and that the soft X-ray excess (i.e., rise in the X-ray spectrum with decreasing frequency) is the high-energy Comptonized (Wien) tail of this distribution. Data courtesy of P. M. Gondhalekar.

has been clearly detected in AGN X-ray spectra at ~ 6.4 keV. The observed equivalent widths of these lines are generally in the range 50–300 keV (Mushotzky, Done, and Pounds 1993). The exact rest energy at which the line is located depends on the ionization state of Fe, and the proximity of the mean energy to the value expected for neutral iron suggests that it arises in low-ionization (cold) material. The observed widths of the Kα line often correspond to velocities of order tens of thousands of kilometers per second (Mushotzky *et al.* 1995). In one extreme case, MCG-6-30-15, the Kα line has a width of $100\,000$ km s^{-1}, and is apparently gravitationally redshifted, which argues strongly for an origin near a supermassive black hole (Tanaka *et al.* 1995).

A simple model that produces many of the features seen in AGN X-ray spectra consists of a cold, thick slab of gas that is illuminated from above (perhaps from the hot corona) with an X-ray power-law spectrum typical of AGNs. The predicted spectrum of such a model is shown in Fig. 4.4. At $h\nu \gtrsim 10$ keV, the Compton reflection spectrum can be seen. This feature peaks in the range 30–50 keV. The reflection model also produces strong Fe Kα emission, and weaker lines as well; however, in this model,

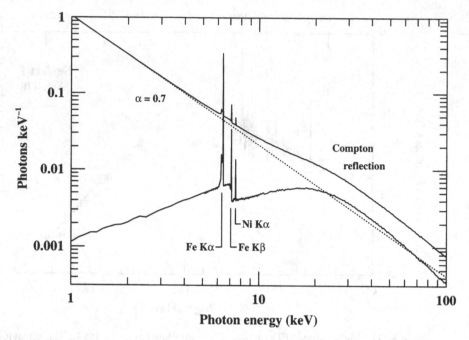

Fig. 4.4. An X-ray 'reflection' model (George and Fabian 1991). An input power-law continuum with energy index $\alpha = 0.7$, shown as a dotted line, irradiates a plane parallel slab of 'cold' (i.e., $T \lesssim 10^6$ K) gas. The lower spectrum shows the reflected X-ray spectrum, which contains the Fe Kα (6.4 keV), Fe Kβ (7.06 keV), and Ni Kα (7.48 keV) fluorescence lines, along with the Fe K-shell absorption edge at ~ 7.11 keV. (The corresponding Ni K-shell absorption edge at 8.34 keV is too weak to be visible.) The effects of Compton 'down-scattering' within the slab (whereby photons interact with and recoil off lower-energy electrons, losing energy in the process) can be seen in the fall-off in the reflected continuum at high energies, in the smearing of the Fe K-shell absorption edge, and in the lower-energy wings of the fluorescence lines (which is most easily seen in the Fe Kα line). The combination of Compton down-scattering of higher-energy photons on the one hand and photoelectric absorption of lower-energy photons on the other gives rise to a broad hump in the 'reflected' continuum over the range ~ 20–100 keV. Data courtesy of I. M. George.

the reflecting slab must cover half of the sky as seen from the X-ray source in order to produce the observed equivalent width of the Kα line. Krolik, Madau, and Życki (1994) point out that at least some of the reflection component might arise at the inner face of the thick obscuring torus that appears to surround the AGN itself and which forms much of the basis for AGN unification models (Chapter 7).

AGNs are strongly variable at X-ray energies (reviewed by McHardy 1988). Rapid X-ray variations are of particular relevance as they may well provide a probe of the innermost regions of the accretion disk. Over long time scales (i.e., days and longer), the variations in the X-ray continuum are apparently correlated with variations at UV/optical wavelengths, although at the present time there are insufficient data to determine whether the X-ray and UV/optical continuum variations are strictly simultaneous, or if the variations in one band precede those in the other, which would provide the direction of the causal relationship. The existing data suggest that the variations are simultaneous to within a few days or less.

As in the case of UV/optical variations, there is no evidence for strictly periodic AGN X-ray variations that might be indicative of orbital motion. One particular case of interest in this regard is the Seyfert 1 galaxy NGC 6814, for which a 3.4-hr period in the X-ray flux has been reported (Mittaz and Branduardi-Raymont 1989), and which subsequently generated a great deal of theoretical activity. However, the X-ray periodicity has now been shown to be due to a cataclysmic variable in the same field as the Seyfert galaxy (Madejski *et al.* 1993).

Short time-scale X-ray variability has been measured in a number of Seyfert 1 galaxies. Key points are:

- On short time scales, the X-ray variability power spectrum has a form $1/f^{1-2}$, i.e., there is relatively more variability at higher temporal frequencies than is seen in the UV/optical. The data on short time-scale variability are very limited, however, and consist primarily of a few 'long-looks' of up to nearly 3×10^5 sec with the *European X-Ray Observatory Satellite* (*EXOSAT*).

- The variability power spectra do not show any rapid steepening at the highest observed frequencies (corresponding to a few hundred seconds), which is what would be expected if there were a minimum time scale set by light-travel time across the X-ray emitting region, which is presumably only a few times R_S (eq. 3.20).

The slope of the power spectrum is steep enough that the total power does not diverge at high temporal frequencies, so total power considerations alone do not require a steep roll-over at high frequencies. In the accretion-disk models, the highest frequencies at which significant power is expected correspond to the light-travel time across ~ 5–$10 R_S$, i.e., $t \approx 50$–$100 (M/10^6 M_\odot)$ sec. For most of the variable AGNs observed to date, the most rapid observable time scales are somewhat longer than this, so there is no clear conflict. Indeed, if the power spectrum is sufficiently steep, no further turn over is required, and no minimum time scale for variations will necessarily be found.

There are very few data available on X-ray variability on long time scales. McHardy (1990) has estimated the power spectrum of the X-ray variations in the Seyfert 2 galaxy NGC 5506 over a broad frequency range and finds that the $1/f$ spectrum found at high frequencies ($\gtrsim 10^{-5}$ Hz) becomes approximately flat, i.e., $P(f) \propto f^0$ (a 'white-noise' power spectrum), at low frequencies ($\lesssim 10^{-7}$ Hz).

The apparent coordination of the X-ray and UV/optical continua and the fact that the power-density spectra of the UV/optical variations are steeper (i.e., with much of the high-frequency variability apparently removed) at least suggest that the UV/optical spectrum could be a reprocessed version of the X-ray spectrum.

4.3   **Infrared Continuum**

The origin of the infrared continuum in different types of AGNs is still somewhat controversial. There are some lower-luminosity objects, a few Seyfert 2 galaxies in particular, in which the IR emission is quite clearly thermal in origin, and dust absorption (e.g., the silicate feature at $10 \, \mu$m) and emission (e.g., the silicate feature at $19 \, \mu$m) features are detected, and in some nearby AGNs the IR continuum source is spatially resolved. However, the nature of the IR continuum in most radio-quiet QSOs and Seyfert galaxies is less certain. Until fairly recently, most of the IR continuum emission has been thought to be the short-wavelength extension of the power-law spectrum that is observed at radio wavelengths, probably generated by the SSC mechanism (Jones, O'Dell, and Stein 1974). There had been suggestions, however, that much or all of the IR continuum was thermal emission from dust (e.g., Rees *et al.* 1969, Rieke 1978, Lebofsky and Rieke 1980). There is a growing, but not complete, consensus that virtually all of the IR continuum in non-blazar AGNs is thermal in origin. Three strong pieces of evidence support the thermal interpretation:

The 1 μm minimum: The existence of the IR bump longward of $1 \, \mu$m has led many authors to conclude that this emission must be thermal, as the required temperatures are in the right range ($\lesssim 2000 \, $K) for hot dust in the nuclear regions. Sanders *et al.* (1989) have shown that a minimum in the SED at $\sim 1 \, \mu$m is a general feature of AGNs. The hottest dust has a temperature of $\sim 2000 \, $K; at higher temperatures, dust grains sublimate. This temperature limit affords a natural explanation for the constancy of the frequency where the near-IR spectrum is weakest, i.e., at the Wien cut-off of a $2000 \, $K blackbody.

We define a 'sublimation radius' as the minimum distance from the AGN at which grains of a given composition can exist. The dust grains closest to an AGN probably are graphite rather than silicate, as graphite has a higher sublimation temperature, and a 'normal' mixture of small graphite and silicate grains as in our own Galaxy seems ruled out by the absence of expected strong emission features (Laor and Draine 1993). The sublimation radius for graphite grains is

$$r = 1.3 \, L_{\mathrm{uv46}}^{1/2} \, T_{1500}^{-2.8} \ \mathrm{pc}, \tag{4.15}$$

where L_{uv46} is the central source UV luminosity in units of 10^{46} ergs s^{-1}, and T_{1500} is the grain sublimation temperature in units of $1500 \, $K (Barvainis 1987).†

IR continuum variability: Clear evidence that the hot-dust scenario for the origin of the IR continuum has some merit is provided by the IR continuum variability characteristics. In §4.1, it was noted that the UV and optical continua vary without a measurable time delay between them, and furthermore the X-ray variability seems to

---

† The $T^{-2.8}$ dependence comes from the wavelength dependence of the absorption efficiency for graphite grains.

be correlated with the UV/optical variability with little if any time delay. The situation with the IR continuum is considerably different, in that the IR continuum shows the same variations as the UV/optical continuum, but with a significant time delay. This is interpreted as a light travel-time effect which occurs because of the separation between the UV/optical and IR continuum-emitting regions; whereas the UV/optical emission arises in a very compact region, the IR emission arises in dust that is far away from the central source. The variations occur as the emissivity of the dust changes in response to the UV/optical continuum that heats it. Probably the best-studied case is Fairall 9 (Clavel, Wamsteker, and Glass 1989). During a remarkable fading of the continuum in this source, the near-IR continuum variations were found to lag behind the UV continuum variations by ~ 400 days. The IR continuum thus arises in a region about 400 light days ($\sim 10^{18}$ cm) from the central source (see §5.5). The UV luminosity of Fairall 9 is $\sim 1.8 \times 10^{46}$ ergs s^{-1}, and inserting this value in eq. (4.15) with the sublimation temperature taken to be 1800 K gives $r \approx 1$ pc $\approx 3 \times 10^{18}$ cm, which is in reasonably good agreement with the observations. For reasonable assumptions about the nature of the graphite grains (size $a \approx 0.05\,\mu$m, mass density $\rho \approx 2.26\,\mathrm{g\,cm}^{-3}$), the total mass of dust needed to produce the observed emission is only $0.02\,M_\odot$. The picture that emerges is that within the sublimation radius, dust is destroyed. Farther out, however, it survives and is heated by the UV/optical radiation from the central source to approximately the equilibrium blackbody temperature. The IR continuum arises as this energy is re-radiated by the dust. In the far-IR, the only AGNs that are found to vary are radio-loud sources (Edelson and Malkan 1987) which also suggests a thermal origin for the far-IR continuum in radio-quiet objects.

The submillimeter break: Observations of the far-IR to submillimeter portion of AGN spectra have been made in a limited number of cases (e.g., Edelson *et al.* 1988, Chini, Kreysa, and Biermann 1989, Hughes *et al.* 1993). These observations show that the submillimeter SED decreases rather sharply as one goes to longer wavelengths, so abruptly that in at least a few cases the spectral index longward of the submillimeter break must be less than the value of -2.5 expected in the case of a synchrotron self-absorbed spectrum (i.e., $F_\nu \propto \nu^{5/2}$). At these long wavelengths, a thermal spectrum can produce a cut-off this sharp because the emitting efficiency of small grains is a sensitive function of frequency, $Q_\nu \propto \nu^\gamma$, typically with $\gamma \approx 2$ (Draine and Lee 1984) so the emitted spectrum can have a very strong frequency dependence, $F_\nu \propto \nu^{2+\gamma}$.

4.4 Radio Continuum

Certainly the one part of the spectrum that is believed *not* to have a thermal origin is the radio spectrum. While historically radio emission was a key factor in the discovery and identification of AGNs, even in radio-loud QSOs the radio spectrum contributes little to the bolometric luminosity. The energy per decade of frequency is about 3 orders of magnitude down from the UV/optical in radio-loud AGNs, but it is lower by 5–6 orders of magnitude in radio-quiet objects. Nevertheless, studies of radio sources are important as a probe of particle acceleration processes near the central sources.

4.4.1   Compact Radio Sources

The radio spectra of the central compact sources in AGNs are flat, and thus the radiation is generally taken to be non-thermal in origin, with a synchrotron mechanism usually invoked. Evidence that the emission is non-thermal is at least twofold:

Spectral shape: The spectral indices for compact sources are close to flat but progressively steeper at shorter wavelengths, which is characteristic of optically thick sources that undergo continued injection of higher-energy electrons. The flatness of the radio spectrum is attributed to a complex source structure, where the low-frequency cut-off is different for different regions of the source (which is what will happen if there are radial gradients in magnetic field and particle density within the source). Low-frequency cut-offs, attributable to synchrotron self-absorption, are detected in some sources, but frequency dependence is usually not as strong as the expected $F_v \propto v^{5/2}$ spectrum, again probably because the source structure is complex, with different regions becoming optically thick at different frequencies.

Brightness temperature: The specific intensity of a radio source at a given frequency can be determined by measuring the flux and angular size of the source. An equivalent temperature can be computed, which tells what the temperature of the source would have to be if it were indeed radiating like a blackbody. At radio wavelengths, we are in the Rayleigh–Jeans limit for any temperature realizable in the Universe (i.e., $T \gtrsim 3\,\mathrm{K}$), so for an optically thick thermal source the intensity is given by the Planck function with $hv \ll kT$,

$$I_v = \frac{F_v}{\pi\theta^2} = B_v = \frac{2kT_\mathrm{B}}{\lambda^2}, \tag{4.16}$$

where F_v is the observed flux at wavelength $\lambda = c/v$ and θ is the angular radius of the source. The quantity T_B is called the 'brightness temperature' of the source, and it equals the kinetic temperature for an optically thick thermal source†. For optically thin thermal emission, the brightness temperature is less than the kinetic temperature. Measurements of F_v and θ for extragalactic compact sources consistently yield peak brightness temperatures in the range 10^{11}–$10^{12}\,\mathrm{K}$, which clearly points to a non-thermal origin for the radio emission.

The peak brightness temperatures of compact sources are always very high, but rarely exceed $T_\mathrm{B} \approx 10^{12}\,\mathrm{K}$, except in the case of blazars (§4.5). There is a fundamental reason

---

† Note that brightness temperature is a function of frequency.

for this that has to do with the energy density of the magnetic field $U_{mag} = B^2/8\pi$, which controls the rate of synchrotron energy losses, relative to the radiation energy density $U_{rad} = 4\pi J/c$, which controls the rate of inverse-Compton scattering. When U_{rad} is increased by synchrotron radiation to the point where it exceeds U_{mag}, then the electrons begin to inverse-Compton scatter off the synchrotron-emitted photons, thus further increasing the photon energies at a rate $\sim \gamma^4 U_{rad}$, where γ is the Lorentz factor for the electrons (i.e., their total energy per particle is $\gamma m_e c^2$). This process very quickly runs away, with the synchrotron photons inverse-Compton scattered up to higher energies, until the onset of the Klein–Nishina cross-section (and energy conservation) terminates the up-scattering process and prevents a catastrophic runaway. The fact that the compact sources are not extraordinarily strong γ-ray sources thus indicates that $U_{rad}/U_{mag} < 1$ in these sources. The upper limit to the synchrotron brightness is imposed by the onset of inverse-Compton scattering, and the brightness temperature of such a source is $\sim 10^{12}$ K over a rather wide range of physical conditions (Kellermann and Pauliny-Toth 1969).

4.4.2 Superluminal Motion

Flat ($\alpha \approx 0$) radio spectra and rapid flux variability suggest that compact radio sources should have structure on small scales. This provided much of the stimulus for the development of very long-baseline interferometry (VLBI), which can provide angular resolution at the milliarcsecond level. VLBI observations of compact sources in AGNs indeed have established in some cases the presence of multiple components and complex structures. These are generally seen in sources with very bright central cores. Repeated observations of these multiple sources sometimes show proper motions of one or more of these sources, always in the sense that they are getting farther apart (e.g., Fig. 4.5). The projected source separations are typically tens of parsecs, and the proper motions observed over a number of years imply transverse velocities as large as $v_T \approx 10c$! In well-observed sources, it is possible to extrapolate the proper motions backward in time, and the time of zero separation between the components is often found to coincide with an outburst of the continuum that is detected over a broad frequency range.

Many explanations have been put forth to explain the existence of these apparently 'superluminal' (i.e., $v > c$) motions in radio sources, including non-cosmological redshifts† and real tachyonic motions. However, the only explanation that does not seem to create more problems than it solves is that these apparent superluminal velocities are attributable to bulk relativistic motion along the line of sight to the continuum source. To see how this works, consider the following model, originally due to Blandford, McKee, and Rees (1977), which is diagrammed in Fig. 4.6.

Consider two sources initially at point B at time t_1, which are detected by an

---

† By supposing that quasars are at distances smaller than implied by their emission-line redshifts and the Hubble law, the observed proper motions reduce to smaller transverse velocities, and the inferred luminosities of quasars are similarly reduced.

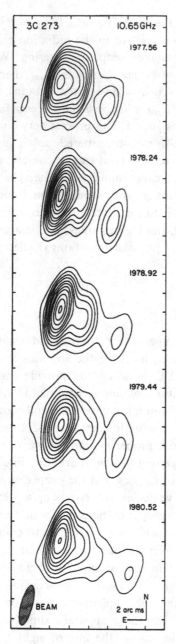

Fig. 4.5. VLBI maps at a frequency of 10.65 GHz of the quasar 3C 273 ($z = 0.158$) made at five different epochs, showing apparent 'superluminal' (i.e., faster than light) expansion of the source (from Pearson *et al.* 1981). Figure courtesy of T. J. Pearson and the California Institute of Technology. Reproduced by permission from *Nature*, Vol. 290, pp. 365–367. Copyright 1981 Macmillan Journals Limited.

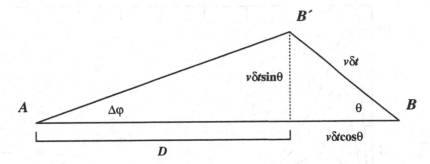

Fig. 4.6. The geometry assumed to explain apparent superluminal expansion of radio sources, i.e., where the observer at A sees a radio source move from B to B' at a speed apparently exceeding the speed of light. This effect can occur as $v \to c$ if the angle to the line of sight θ is small, but non-zero.

observer at point A at time t_1' after the light has traversed the distance AB. At some later time $t_2 = t_1 + \delta t$, one of the sources has moved a distance $v\delta t$. Observations of the two sources are again made, which the observer again records at some later time t_2' on account of the light-travel time between the source and the observer. The distance between A and B is $D + v\delta t\cos\theta$; however, owing to the motion of one of the sources in the radial direction, the distance between the observer at A and the moving source at B' is approximately D. The angular separation between the two sources at the second observation is

$$\Delta\phi = \frac{v\delta t \sin\theta}{D}. \tag{4.17}$$

The time that the observer measures between the two observations can be computed by noting

$$t_1' = t_1 + \frac{D + v\delta t\cos\theta}{c} \tag{4.18}$$

and

$$t_2' = t_2 + \frac{D}{c}, \tag{4.19}$$

so that the measured interval between the observations is

$$\begin{aligned} \Delta t &= t_2' - t_1' = t_2 - t_1 - \frac{v\delta t\cos\theta}{c} \\ &= \delta t\,(1 - \beta\cos\theta), \end{aligned} \tag{4.20}$$

where $\beta = v/c$, in the usual convention. Thus, the transverse velocity inferred by the observer is

$$\beta_{\rm T} = \frac{v_{\rm T}}{c} = \frac{D}{c}\frac{\Delta\phi}{\Delta t} = \frac{v\sin\theta}{c(1 - \beta\cos\theta)} = \frac{\beta\sin\theta}{1 - \beta\cos\theta}. \tag{4.21}$$

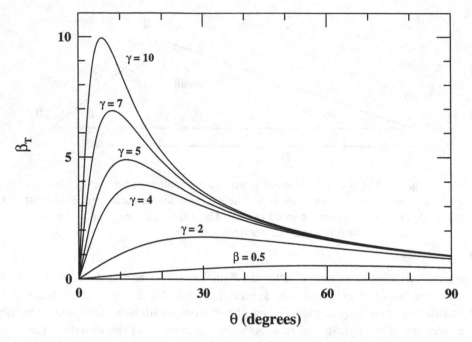

Fig. 4.7. The apparent transverse velocity $\beta_T = v_T/c$ of a source moving at an angle θ to the observer's line of sight (as in Fig. 4.5) as a function of the Lorentz factor $\gamma = (1 - \beta^2)^{-1/2}$. For $\gamma \gg 1$, the apparent transverse velocity can exceed the speed of light.

In Fig. 4.7, we show the measured transverse velocity β_T as a function of the angle to the line of sight θ for various values of the familiar Lorentz factor, $\gamma = (1 - \beta^2)^{-1/2}$.

At what value of θ is β_T maximized? This can be found by differentiating eq. (4.21) with respect to θ and setting the result equal to zero, i.e.,

$$\frac{\partial \beta_T}{\partial \theta} = \frac{\beta \cos \theta}{1 - \beta \cos \theta} - \frac{(\beta \sin \theta)(\beta \sin \theta)}{(1 - \beta \cos \theta)^2} = 0, \tag{4.22}$$

so that

$$\beta \cos \theta (1 - \beta \cos \theta) = \beta^2 \sin^2 \theta = \beta^2 - \beta^2 \cos^2 \theta, \tag{4.23}$$

which yields $\theta_{max} = \cos^{-1} \beta$. By inserting this into eq. (4.21), and noting that

$$\sin(\cos^{-1} \beta) = (1 - \beta^2)^{1/2}, \tag{4.24}$$

we obtain

$$\beta_T^{max} = \frac{\beta (1 - \beta^2)^{1/2}}{(1 - \beta^2)} = \beta \gamma. \tag{4.25}$$

Thus, as $\beta \to 1$, $\beta_T^{max} \approx \gamma$, which can be arbitrarily high. For even modest relativistic bulk motions close to the line of sight (i.e., $\theta \approx \cos^{-1} \beta$) the observer may detect projected transverse velocities apparently in excess of the speed of light, as shown in Fig. 4.7, of order γc.

4.5 Blazar Spectra

Blazars form a small subset of AGNs, constituting less than a few percent of the known AGN population. While blazars have some properties in common with other AGNs, they have several characteristics that set them apart. Blazars share the following characteristics (Miller 1989, Bregman 1990):

- A smooth (i.e., featureless) continuum that originates in an unresolved source.
- Rapid large-amplitude optical variability (on time scales of days or shorter) in both flux and polarization.
- Highly variable optical polarization that can reach rather large values ($\gtrsim 3$–4%).
- Strong and variable radio emission.

Blazars can be further subdivided into BL Lac objects (first recognized as a distinct class of object by Strittmatter *et al.* 1972) and optically violent variables (OVVs, which were first recognized as distinct from other quasars by Penston and Cannon 1970), as mentioned in Chapter 2. BL Lac objects have emission and absorption lines that are weak relative to the continuum (i.e., small equivalent widths). Furthermore, when lines are detectable, the redshifts tend to be small ($z \lesssim 0.1$). In OVVs on the other hand, broad emission lines are prominent spectral features, except when the continuum is at its brightest, when the emission-line equivalent widths become small as the line flux changes little or not at all while the continuum increases dramatically. The redshifts of OVVs tend to be large ($z \gtrsim 0.5$).

All known blazars are radio-loud, core-dominated sources; no radio-quiet blazars have been found, although they have been searched for (Stocke *et al.* 1990). The IR brightness temperatures of blazars are sometimes found to be greater than 10^6 K, and the radio brightness temperatures of these sources are in excess of 10^{12} K, in some cases greater than 10^{19} K (Quirrenbach *et al.* 1991). Some blazars are superluminal sources; the apparent transverse velocities are modest for superluminal BL Lac objects (typically $\beta_T \approx 3h_0^{-1}$), whereas the distribution of β_T for superluminal quasars is approximately flat up to $\beta_T \approx 10h_0^{-1}$ (Cohen 1989).

Typical blazar SEDs are shown in Fig. 4.8. An important feature of blazar SEDs that is immediately obvious is that the peak of the SEDs occurs at different wavelengths for different sources. Some blazars emit most of their energy in the IR, others in the UV/optical. In some blazars, the SEDs drop very rapidly at high energies. In other cases (e.g., Mrk 421), they extend to very high energies. As noted earlier (§4.2), nearly all AGNs detected at MeV energies and above are blazars (Maisack *et al.* 1995, von

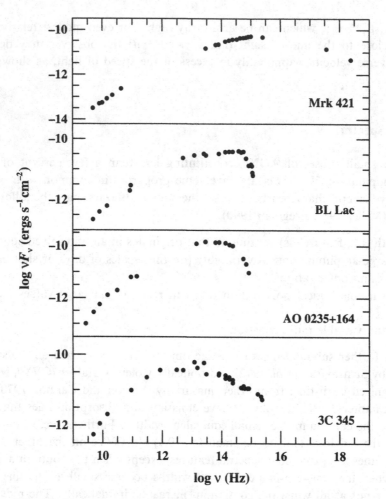

Fig. 4.8. Spectral energy distributions for four blazars. The peak in the SED is at progressively lower photon energies going from top to bottom. There are no identifiable features at particular energies, as there are in non-blazar AGN SEDs (Figs. 1.3 and 4.1), which is consistent with a non-thermal origin for the blazar spectra. The sharp high-energy cut-offs in BL Lac and AO 0235+164 are thought to be due to corresponding high-energy cut-offs in the electron distribution. The data are from the following sources – Mrk 421: Makino *et al.* (1987); BL Lac: Bregman *et al.* (1990); AO 0235+164: a compilation from the literature provided courtesy of J. N. Bregman (see also Bregman 1990); and 3C 345: Bregman *et al.* (1986). After a similar diagram by Bregman (1990).

Montigny *et al.* 1995). At least one blazar, Mrk 421, has been detected at TeV energies (i.e., 10^{12} eV or $v \approx 10^{26}$ Hz) via atmospheric Čerenkov imaging (Punch *et al.* 1992), but attempts to detect other AGNs in this way have yielded only upper limits (Kerrick *et al.* 1995). The characteristic features of AGN SEDs listed at the beginning of this

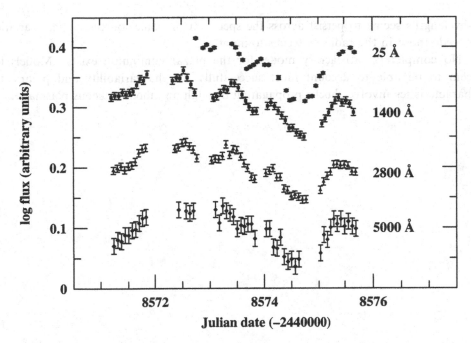

Fig. 4.9. Light curves of the BL Lac object PKS 2155 − 304 at X-ray, UV, and optical wavelengths, obtained during a multiwavelength monitoring campaign in 1991 November (from Edelson *et al.* 1995). The UV/optical variations appear to follow the X-ray variations with a time delay of 2–3 hours. The fluxes are log F_ν, arbitrarily normalized for display on a common scale. Data courtesy of R. Edelson.

chapter (the big blue bump, the 1 μm minimum, the submillimeter break) are missing in blazar spectra. The absence of spectral features at characteristic photon energies effectively precludes a thermal origin for the continuum.

Blazar variability provides important insights in regard to the origin of the continuum. The strength of the continuum variations are correlated with both the apparent luminosity (i.e., assuming isotropic emission) and degree of continuum polarization, which argues strongly for a non-thermal origin (Edelson 1992). The optical fluxes of blazars can range over several magnitudes on time scales of months to years (e.g., Webb *et al.* 1988). 'Microvariability', or smaller-amplitude rapid variability (e.g., 0.1 mag changes on time scales as short as hours), has been confirmed in some blazars (Miller, Carini, and Goodrich 1989). Recent multiwavelength spectral monitoring programs have begun yielding results that greatly constrain models. For example, the UV continuum (1400 Å) variations in the well-observed BL Lac object PKS 2155−304 (Fig. 4.9) follow the X-ray (25 Å) variations with a delay of about 2–3 hr; this rules out the SSC mechanism, in which the inverse-Compton X-ray photons would be expected to *lag behind* rather than precede the long-wavelength synchrotron seed photons. The general pattern of variations occurring first at short wavelengths and later at longer wavelengths seems to persist across the spectrum; in some sources, optical variations precede those in the radio by weeks to months.

No completely satisfactory model for the blazar continuum exists. Models that seem to be able to account most successfully for the variability and polarization characteristics involve shocks propagating through an inhomogeneous plasma jet.

5 The Broad-Line Region

Broad emission lines are one of the dominant features of many AGN spectra. The broad-line region (BLR) plays a particularly important role in our understanding of AGNs by virtue of its proximity to the central source. The BLR potentially can provide a useful probe of the central source for at least two reasons. First, the bulk motions in the BLR are almost certainly determined by the central source, with gravity and radiation pressure competing. Second, the BLR reprocesses energy emitted by the continuum source at ionizing ultraviolet energies that cannot be observed directly, and thus the emission lines provide indirect information about this important part of the continuum.

5.1 Broad-Line Spectra

The broad-line spectra of AGNs show a great deal of diversity in terms of relative line strengths and profiles. The spectra shown in Figs. 1.1, 1.2, and 2.2 should be understood to be only more or less typical, as are the line strengths and equivalent widths of the stronger lines given in Table 1.1.

Widths of the broad lines show considerable differences from object to object. It is assumed that the lines are Doppler-broadened (§5.2), so line widths are almost always measured in velocity units, usually in terms of either (a) the full width at half maximum intensity (FWHM) or (b) the full width at zero intensity (FWZI); while the latter better reflects the true range of line-of-sight velocities, it is subject to uncertainties arising from ambiguities as to where the line actually reaches the continuum level and, in cases of especially high velocities, blending of the wings of different lines. The widths of AGN broad lines range from a minimum of $\Delta v_{FWHM} \approx 500\,\mathrm{km\,s^{-1}}$ (only somewhat broader than the narrow lines) to $\Delta v_{FWHM} \gtrsim 10^4\,\mathrm{km\,s^{-1}}$, with typical values $\Delta v_{FWHM} \approx 5000\,\mathrm{km\,s^{-1}}$; in some cases, the width at zero intensity can exceed $30\,000\,\mathrm{km\,s^{-1}}$. It is also notable that even in a single spectrum different emission lines might have different widths. It is often found in optical spectra, for example, that the helium lines He II $\lambda 4686$ and He I $\lambda 5876$ are broader than the hydrogen Balmer lines at any given relative intensity level.

The line profiles are also diverse. To a low-order approximation, AGN emission-line profiles are often described as 'logarithmic', i.e., the flux at radial velocity Δv from line center is $F_\lambda(\Delta v) \propto -\ln \Delta v$ (§5.4). As shown in Fig. 5.1, this is indeed a reasonable description in some cases, but in other cases the line profiles show fairly complicated and often variable structure. Again, the line structure can vary considerably among different lines in a given spectrum. From this alone, we can immediately infer that the relative line strengths (e.g., the Balmer decrement) vary as a function of line-of-sight velocity, which requires some sort of structured BLR.

In addition to the strong lines listed in Table 1.1, a number of other lines are also often detected in AGNs. In many cases, the large Doppler widths cause spectral features to be severely blended; a list of strong blended features is given in Table 5.1. Many of the strongest features (e.g., C IV λ1549) are seen to be doublets with such small separations that they are unresolved in AGN emission spectra. The helium lines He II λ4686 and He I λ5876 are usually present although they are sometimes hard to isolate because they are low-contrast features (low peak intensities and large widths). Moreover, He II λ4686 is usually blended with a very broad feature that covers most of the region between Hγ and Hβ and is identified as originating in a large number of weak blended lines of Fe II. A similar Fe II blend is seen at wavelengths immediately longward of Hβ (see Fig. 2.2); following Osterbrock (1977), we label the complex Fe II blends shortward and longward of Hβ as Fe II λ4570 and Fe II $\lambda\lambda$5190, 5320, respectively. Finally, there is a very strong blended feature that consists of blended Fe II emission lines and Balmer continuum emission that extends from ~ 2000 Å to ~ 4000 Å (Wills, Netzer, and Wills 1985). This feature, the 'small blue bump', is shown in Fig. 5.2.

In objects of sufficiently high redshift, additional far-UV lines such as O VI λ1035 are detected. In the IR, hydrogen Paschen-series lines and other emission features are also seen.

Typical values of line flux ratios, e.g., Lyα/H$\beta \approx 5$–15 and Hα/H$\beta \approx 4$–6 differ significantly from their Menzel–Baker Case B recombination values† (i.e., Lyα/H$\beta \gtrsim 30$ and Hα/H$\beta \approx 2.8$); part of the discrepancy can be attributed to reddening within our own Galaxy and within the vicinity of the AGN (typically $E(B - V) \lesssim 0.3$), but radiative-transfer effects (resulting from non-negligible optical depth in the $n \geq 2$ series lines because of collisional excitation of the excited states) in the BLR cause significant deviations from Case B recombination in this relatively high-density nebular environment.

Emission-line fluxes vary with time in a way that is highly correlated with the continuum fluxes, which argues very convincingly that photoionization by the central source drives the emission lines and that a large fraction of the recombination emission

---

† Baker and Menzel (1938) considered the emitted spectrum due to recombination in a pure hydrogen nebula. In 'Case A', all of the lines are assumed to be optically thin, so that each radiative transition produces a photon that escapes from the nebula. In 'Case B', it is assumed that the Lyman-series lines are all optically thick; in this case, Lyman-series photons produced by radiative transitions to the ground state are immediately reabsorbed. Lyα photons gradually diffuse out of the nebula, usually by diffusion in velocity space. Through successive emissions and absorptions, higher-order Lyman lines are eventually degraded to lower-energy photons in other series (which escape) plus a Lyα photon; e.g., a Lyγ ($n = 3 \rightarrow n = 1$) photon becomes an Hα ($n = 3 \rightarrow n = 2$) photon plus a Lyα ($n = 2 \rightarrow n = 1$) photon. The Case B predictions for relative strengths of the $n \geq 2$ series lines are a good first approximation to the observed values in low-density nebulae. A more complete discussion is given by Osterbrock (1989), pp. 77–86.

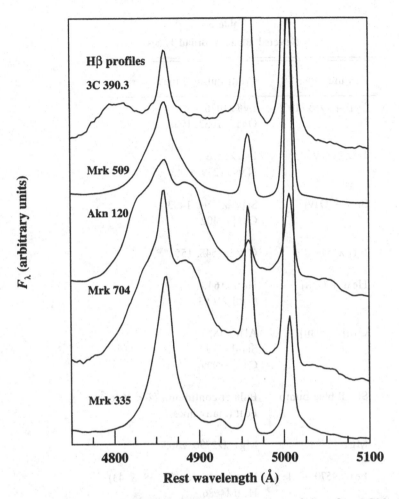

Fig. 5.1. The Hβ emission-line profiles in the spectra of these five AGNs give an indication of the diversity seen in broad-line profiles. Each of these spectra represents an average of a large number of individual spectra obtained over many months; while the total fluxes in the emission lines vary on short time scales, the features in the line profiles persist for at least months or years. The source 3C 390.3, which has a double-peaked profile (the red peak is somewhat obscured by blending with the [O III] λ4959 line), is a broad-line radio galaxy, and the other sources are all Seyfert 1 galaxies. These spectra have all been shifted to the rest frame of the AGN, and the flux scale is arbitrary. The resolution of these spectra is about 10 Å. All of the data here were obtained with the Ohio State CCD spectrograph as part of an AGN spectroscopic monitoring program.

originates in clouds that are optically thick in the ionizing continuum. The continuum and emission-line variations can be very dramatic. In some cases, Seyfert 1 galaxies have become so faint that the broad components of the emission lines have almost, but not completely, vanished on some occasions.

Table 5.1
Selected Blended Broad Lines

Feature	Contributing Lines
Lyβ + O vi λ1035	Lyβ λ1026 O vi $\lambda\lambda$1032, 1038
Lyα + N v	Lyα λ1216 N v $\lambda\lambda$1239, 1243
Si iv + O iv]	Si iv $\lambda\lambda$1394, 1403 O iv] λ1402
C iv λ1549	C iv $\lambda\lambda$1548, 1551
He ii + O iii]	He ii λ1640 O iii] λ1663
C iii] + Si iii]	Al iii λ1857 Si iii] λ1892 C iii] λ1909
Small blue bump	Balmer continuum ($\lambda < 3646$ Å) Fe ii (many lines)
Mg ii λ2798	Mg ii $\lambda\lambda$2796, 2803
Fe ii λ4570 + He ii	Fe ii (multiplets 37, 38, & 43) He ii λ4686
Hβ	Hβ λ4861 Fe ii $\lambda\lambda$4924, 5018 (multiplet 42)
Fe ii $\lambda\lambda$5190, 5320	Fe ii (multiplets 42, 48, 49, and 55)

5.2   Basic Parameters

In contrast to normal H ii regions, planetary nebulae, and AGN narrow-line regions (Chapter 6), there are no simple temperature and density diagnostics for AGN broad-line spectra. The fundamental reason for this is that the BLR electron densities are sufficiently high that virtually all forbidden lines are collisionally suppressed and the emissivity j (ergs s^{-1} cm^{-3} ster^{-1}) of all these lines is in the high-density limit where $j \propto n$ (see §6.2.1). Nevertheless, the similarity of relative line intensities to those in other

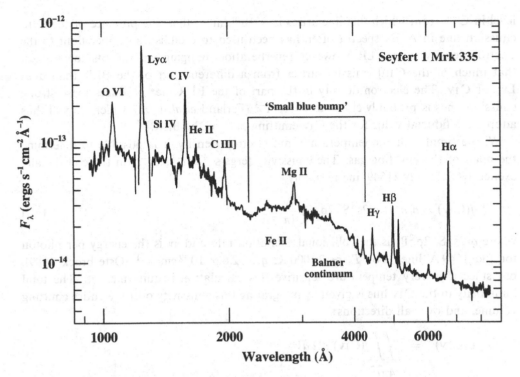

Fig. 5.2. The 'small blue bump', a blend of Balmer continuum and Fe II line emission is prominent in the Seyfert 1 galaxy Mrk 335 (Zheng *et al.* 1995a) and shows up particularly well on a log–log plot. Other strong features are also labeled. Data courtesy of W. Zheng.

ionized gases indicates that the gas temperature is of order 10^4 K. The line-of-sight velocity dispersion for a 10^4 K gas is

$$v \approx \left(\frac{kT}{m_{\rm p}}\right)^{1/2} \approx 10 \text{ km s}^{-1}, \tag{5.1}$$

but as noted earlier typical broad-line widths are ~ 5000 km s^{-1}. If the broad-line widths reflect purely thermal motions, the gas temperature inferred from eq. (5.1) would be $T \gtrsim 10^9$ K! Clearly some other broadening mechanism is required, and this is usually attributed to differential Doppler shifts due to bulk motions of individual line-emitting clouds.

The [O III] $\lambda\lambda 4363, 4959, 5007$ lines are usually strong in highly ionized gases, but they appear to be absent or at least extremely weak in broad-line spectra. The critical density for collisional de-excitation of the $^1S_0$ level in O^{++} (the upper level for the $\lambda 4363$ transition) is $\sim 10^8$ cm^{-3}, so this provides a lower limit for the electron density in the BLR. The only strong non-permitted line in UV/optical broad-line spectra is the intercombination line C III] $\lambda 1909$. The critical density for de-excitation of the $^3P_1$

level in C^{++} from which this line arises is $\sim 10^{10}$ cm^{-3}; thus, the presence of the $C\,III]$ emission line in AGN spectra often has been used to establish an upper limit to the electron density in the BLR. However, reverberation-mapping experiments now suggest that much of the $C\,III]$ emission arises from a different part of the BLR than does $Ly\alpha$ or $C\,IV$. The electron density in the part of the BLR that produces these strong emission lines is probably closer to 10^{11} cm^{-3} (Ferland *et al.* 1992), which we will thus adopt as a fiducial value for the $C\,IV$-emitting part of the BLR.

With estimates of the temperature and electron density, it is possible to determine the mass of the emitting gas. The emissivity (ergs s^{-1} cm^{-3} ster^{-1}) in the collisionally excited (§6.2.1) $C\,IV\,\lambda1549$ line is simply

$$j(C\,IV) = n_e n_{C^{3+}} q(2s\,^2S, 2p\,^2P) \frac{h\nu}{4\pi} \qquad (5.2)$$

where $q(2s\,^2S, 2p\,^2P)$ is the collisional excitation rate and $h\nu$ is the energy per photon for the 1549 Å line. For $T_e \approx 20\,000$ K, $q \approx 2.6 \times 10^{-9}$ cm^3 s^{-1} (Osterbrock 1989), but since it is very temperature sensitive this calculation is quite inexact. The total luminosity in the $C\,IV$ line is given by integrating this emissivity over the entire emitting volume, and over all directions:

$$
\begin{aligned}
L(C\,IV) &= \int\int j(C\,IV)\, d\Omega\, dV \\
&= \frac{4\pi r^3}{3} \epsilon n_e n_{C^{3+}} q(2s\,^2S, 2p\,^2P) h\nu \\
&\approx \epsilon r^3 n_e^2\, 4.6 \times 10^{-23} \text{ ergs s}^{-1},
\end{aligned}
\qquad (5.3)
$$

where ϵ is a 'filling factor' which indicates how much of the emitting volume actually contains line-emitting material (we can assume that the rest of the volume is a vacuum). For the sake of simplicity, we have assumed a cosmic abundance of carbon (log [C/H] = -3.48) and that all of the carbon is triply ionized. For spherical clouds of radius ℓ in a spherical BLR of radius r, the filling factor is

$$\epsilon = \frac{N_c\left(4\pi\ell^3/3\right)}{\left(4\pi r^3/3\right)} = \frac{N_c\,\ell^3}{r^3}, \qquad (5.4)$$

where N_c is the number of clouds within the BLR. Equation (5.3) can be solved for ϵ, since all of the other parameters can be estimated. To put this in units appropriate for luminous Seyfert galaxies, the $C\,IV$ luminosity is written as

$$L(C\,IV) = L_{42}(C\,IV) \times 10^{42} \text{ ergs s}^{-1}, \qquad (5.5)$$

and

$$r = 8h_0\, L_{42}^{1/2}(C\,IV) \text{ light days} = 2 \times 10^{16}\, h_0\, L_{42}^{1/2}(C\,IV) \text{ cm}, \qquad (5.6)$$

where the BLR size r is measured by reverberation mapping (§5.5). This gives

$$\epsilon = 2.7 \times 10^{-7}\, h_0^{-3}\, L_{42}^{-1/2}(C\,IV), \qquad (5.7)$$

where an electron density $n_e = 10^{11}\,\text{cm}^{-3}$ has been assumed. The small value of the filling factor indicates that the BLR structure is very filamentary or clumpy, and it is then natural to explain the emission-line widths as due to bulk motions of individual clouds, each radiating at its thermal width (eq. 5.1).

The total mass of the BLR also can be estimated. We can combine eqs. (5.3) and (5.5) to obtain

$$\epsilon r^3 = \frac{2.2 \times 10^{64}}{n_e^2} L_{42}(\text{C\,\textsc{iv}}). \tag{5.8}$$

The mass of the BLR is given by the product of the cloud mass density $n_e m_p$, the volume of each cloud $4\pi \ell^3/3$, and the total number of clouds N_c, i.e.,

$$
\begin{aligned}
M_{\text{BLR}} &= \frac{4\pi}{3}\,\ell^3\,N_c\,n_e m_p \\
&= \frac{4\pi}{3}\,\epsilon r^3\,n_e m_p \\
&\approx \frac{1.5 \times 10^{41}}{n_e}\,L_{42}(\text{C\,\textsc{iv})}\,\text{g} \\
&\approx 10^{-3}\,L_{42}(\text{C\,\textsc{iv}})\,M_\odot,
\end{aligned}
\tag{5.9}
$$

where we have used eq. (5.4) and in the last step we have taken $n_e = 10^{11}\,\text{cm}^{-3}$, as before. The mass of the BLR is trivial compared to the mass of the central source;† for even the most luminous AGNs, less than $10\,M_\odot$ of line-emitting gas is required.

Another important quantity to estimate is how much of the continuum emission is actually absorbed by the BLR. The recombination lines vary strongly in response to continuum variations, which indicates that much of the BLR emission arises in clouds that are optically thick in the ionizing continuum. In the case of optically thick clouds, the fraction of the ionizing continuum emission which is absorbed by BLR clouds is simply given by the fraction of the sky covered by BLR clouds as seen at the central source, $f = \Omega/4\pi$, which is known as the 'covering factor'.

One simple way to estimate the covering factor is to recall that in an optically thick pure hydrogen nebula, every photoionization ultimately results in one Lyα photon.‡ If all of the ionizing photons are absorbed, there is a simple relationship between the total Lyα flux and the continuum strength, i.e., the equivalent width of Lyα. Denoting the frequency corresponding to a wavelength of $1216\,\text{Å}$ as ν_0, the continuum flux at $1216\,\text{Å}$ as $F_\nu(1216\,\text{Å})$ $(\text{ergs s}^{-1}\,\text{cm}^{-2}\,\text{Hz}^{-1})$, and approximating the ionizing continuum as a power law of index α_{ox} (eq. 4.10), the flux of ionizing photons emitted by the source $(\text{photons s}^{-1}\,\text{cm}^{-2})$ is given by integrating over all ionizing frequencies (i.e., above ν_1,

---

† This value for the emitting gas is somewhat smaller than given elsewhere in the literature because of the high density we use, since the emissivity is proportional to n_e^2.

‡ This statement remains approximately true even in more sophisticated optically thick models with more realistic elemental abundances (although see Shields and Ferland 1993).

the frequency of the Lyman edge at 912 Å), i.e.,

$$
\begin{aligned}
\Phi(\mathrm{H}) &= \int_{\nu_1}^{\infty} \frac{F_\nu \, d\nu}{h\nu} \\
&= \int_{\nu_1}^{\infty} \frac{F_\nu(1216\,\text{Å})}{h\nu} \left(\frac{\nu}{\nu_0} \right)^{-\alpha_{\mathrm{ox}}} d\nu \\
&= \frac{F_\nu(1216\,\text{Å})\nu_0^{\alpha_{\mathrm{ox}}}}{h} \int_{\nu_1}^{\infty} \nu^{-1-\alpha_{\mathrm{ox}}} \, d\nu \\
&= \frac{F_\nu(1216\,\text{Å})}{h\alpha_{\mathrm{ox}}} \left(\frac{\nu_0}{\nu_1} \right)^{\alpha_{\mathrm{ox}}} \\
&= \frac{F_\nu(1216\,\text{Å})}{h\alpha_{\mathrm{ox}}} \left(\frac{912}{1216} \right)^{\alpha_{\mathrm{ox}}}.
\end{aligned}
\tag{5.10}
$$

Since the number of ionizing photons equals the number of Lyα photons, the flux in the Lyα emission line is

$$
F(\mathrm{Ly}\alpha) = \Phi(\mathrm{H}) \, h\nu_0 = \frac{F_\nu(1216\,\text{Å})}{\alpha_{\mathrm{ox}}} \left(\frac{912}{1216} \right)^{\alpha_{\mathrm{ox}}} \frac{c}{1216\,\text{Å}}
\tag{5.11}
$$

so the equivalent width of the Lyα emission line is simply

$$
W(\mathrm{Ly}\alpha) = \frac{F(\mathrm{Ly}\alpha)}{F_\lambda(1216\,\text{Å})}.
\tag{5.12}
$$

With eqs. (1.6) and (5.11), this becomes

$$
W(\mathrm{Ly}\alpha) = \frac{1216}{\alpha_{\mathrm{ox}}} \left(\frac{912}{1216} \right)^{\alpha_{\mathrm{ox}}} \text{Å}.
\tag{5.13}
$$

For $\alpha_{\mathrm{ox}} = 1.4$ (§4.2), the predicted equivalent width of Lyα is thus 580 Å, if all of the continuum radiation is absorbed. This is about an order of magnitude larger than the observed Lyα equivalent width (Table 1.1), which suggests that only ∼10% of the ionizing continuum radiation is absorbed by the BLR. Thus, the BLR covering factor is probably of order $f \approx 0.1$.

In principle, one might look directly at the continuum at wavelengths just shortward of the Lyman limit in a large number of AGNs to determine the fraction of sources in which no photons are detected shortward of the Lyman limit. This fraction thus equals the mean covering factor for the AGN sample. An equivalent test would be to look for absorption lines that might be produced by a BLR cloud along the line of sight to the continuum source. In practice, such tests have not been carried out successfully, as there are a number of fundamental problems with them. Most importantly, these tests are not very sensitive if the projected sizes of the individual BLR clouds are smaller than the projected size of the continuum source, as indeed we will see is expected to be the case – even if a BLR cloud falls directly along our line of sight to the continuum source, it will only absorb a fraction of the continuum which is given by the ratio of these projected areas. Furthermore, detection of an absorption line or edge requires that the adjacent continuum is optically thin, and this may not be a good assumption. Finally, even if an absorption edge were to be detected, it would not be clear whether it is due to BLR clouds, the Lyman edge in an accretion-disk spectrum, or intervening material, with the latter source becoming progressively more important in higher-redshift QSOs (Chapter 12), where ironically the opportunity to observe the Lyman edge becomes greater.

5.3 Photoionization of the BLR

The energy source that drives the broad emission lines in AGN spectra is almost certainly photoionization by the continuum radiation from the central source, since the emission-line fluxes vary strongly in response to changes in the continuum flux. Because the recombination lines are strongly variable, it can be concluded that a significant fraction of the BLR emission arises in clouds that are optically thick to ionizing photons of energy $h\nu \geq 13.6\,\text{eV}$. There is, however, also some evidence (described below) for a population of BLR clouds in which hydrogen is completely ionized and which thus are optically thin in the Lyman continuum.

The number of photons emitted by the central source per second which can ionize hydrogen is given by

$$Q(\text{H}) = \int_{\nu_1}^{\infty} \frac{L_\nu}{h\nu}\, d\nu, \tag{5.14}$$

where L_ν is the specific luminosity of the central source and the integral is over all hydrogen-ionizing photons.

From this, an 'ionization parameter' U is defined as the ratio of the photon number density to particle density at the incident face of the cloud, i.e.,

$$U = \frac{Q(\text{H})}{4\pi r^2 c\, n_{\text{H}}}. \tag{5.15}$$

Modern photoionization models depend on (a) the shape of the ionizing continuum, (b) the elemental abundances (usually taken to be solar or cosmic), (c) the particle density in the cloud (usually constrained by the presence or absence of various lines, as mentioned earlier), (d) the column density of the cloud (a free parameter, which is sometimes constrained by the observations), and (e) the ionization parameter. The dependence of various line ratios on some of these parameters is shown for a sample photoionization calculation in Fig. 5.3.

Until fairly recently, photoionization models of the BLR have usually been 'single-zone' models, i.e., models in which the line emission of a single cloud is typical of all clouds in the BLR. In these models, the C III]/C IV flux ratio has been used as an indicator of the value of U because it is a sensitive function of U and it is insensitive to elemental abundances. The presence of C III] $\lambda1909$ and absence of [O III] lines was assumed to imply a particle density of order $10^{9.5}\,\text{cm}^{-3}$. However, these values in conjunction with eq. (5.15) imply a BLR size that is larger than the measured

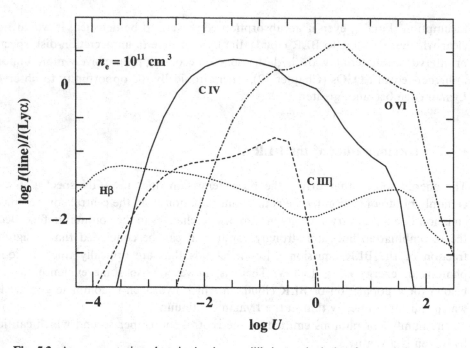

Fig. 5.3. A representative photoionization equilibrium calculation for conditions thought to typify BLR clouds. In these model calculations, the total cloud column density is held fixed at 10^{23} cm^{-3}, and solar elemental abundances and a typical AGN continuum shape are assumed. In this particular model, the electron density is held fixed at 10^{11} cm^{-3} and the ionization parameter U is allowed to vary. The predicted total intensities of the collisionally excited lines C IV λ1549 (solid line), C III] λ1909 (dashed line), O VI λ1035 (dot-dashed line), and the recombination line Hβ λ4861 (dotted line), in each case relative to the total Lyα λ1216 intensity, are shown as a function of U. The ionization structure of the cloud is highly stratified – within the cloud the relative fractions of ions X^{n+1}/X^n decreases with increasing depth into the cloud; at least in the cases of H and He, the transition regions are rather abrupt 'ionization fronts'. Comparison of the relative strengths of C IV and C III] shows how C IV becomes relatively stronger as the ionizing flux (which is proportional to U) increases. The abrupt drop in the C III] intensity at $\log U \approx -0.5$ occurs when throughout the entire cloud carbon is at least triply ionized. This typically occurs at about the same value of U where the entire cloud becomes optically thin in the H-ionizing continuum. Similar effects occur at $\log U \approx 1.5$ when the abundance of C^{3+} becomes negligible, and at $\log U \approx 2$ when the same effect occurs with O^{5+}. Calculations carried out by K. T. Korista and G. J. Ferland using the photoionization equilibrium code CLOUDY.

sizes by about an order of magnitude. More recent models, based on reverberation-mapping results (§5.5), imply that the BLR has a radial ionization structure, and in fact C III] λ1909 emission is collisionally suppressed in the zone which produces most of the C IV emission. There is little one can infer directly about the ionization parameter U or the particle density from the C III]/C IV flux ratio alone; more detailed calculations are required, and these lead to the fiducial values $n_e \approx 10^{11}\,\mathrm{cm}^{-3}$ and $U \approx 0.04\,h_0^{-2}$ in the case of the best-studied Seyfert galaxy, NGC 5548 (Ferland *et al.* 1992).

The requirement that the clouds are optically thick at the Lyman edge allows us to compute a minimum thickness for line-emitting clouds simply by requiring photoionization equilibrium, i.e., over the ionized region, the photoionization rate equals the recombination rate. The number of photoionizations per second in the cloud equals the number of ionizing photons per second incident upon it, i.e., if the cross-section of a BLR cloud is A_c, the number of photons hitting the cloud per second is

$$A_c \frac{Q(\mathrm{H})}{4\pi r^2} = A_c\, U c\, n_e, \qquad (5.16)$$

where eq. (5.15) has been used. The number of recombinations per second per unit volume is $n_e \alpha_B$, where α_B is the Case B hydrogen recombination coefficient ($\alpha_B = 2.52 \times 10^{-13}\,\mathrm{cm}^3\,\mathrm{s}^{-1}$ for $T = 20\,000\,\mathrm{K}$, Osterbrock 1989). Within the ionized zone, the number of photoionizations per second equals the number of recombinations, so

$$A_c\, U c\, n_e = n_e^2\, \alpha_B\, V_c, \qquad (5.17)$$

where V_c is the ionized volume of the cloud. We can identify the Strömgren depth $r_1 = V_c/A_c$, which is thus

$$r_1 = \frac{U c}{n_e \alpha_B}. \qquad (5.18)$$

For our fiducial values for U and n_e,

$$r_1 \approx 4.8 \times 10^{10}\,\mathrm{cm} \approx 0.7 R_\odot. \qquad (5.19)$$

A key characteristic of AGN photoionization models is that they have an extended partially ionized zone (PIZ) which extends far beyond the Strömgren depth within the cloud (e.g., Kwan and Krolik 1981). The existence of the zone is a consequence of a relatively flat ionizing spectrum. The cross-section for ionization of hydrogen falls off rapidly with increasing frequency above the Lyman limit ($\sigma(\mathrm{H}) \propto (\nu/\nu_1)^{-3}$), so the gas absorbs the photons nearest to the edge most efficiently. For a gas ionized by a thermal spectrum, the overwhelming number of ionizing photons lie very close to the Lyman limit in any case, and when these photons are entirely absorbed, the gas quite abruptly becomes neutral. By contrast, with a power-law spectrum, there are still a relatively large number of photons available at higher energies even after the photons near the Lyman limit are all absorbed. This gives rise to a large PIZ, where $\mathrm{H}^+/\mathrm{H}^0 \approx 0.1$. Because Ly$\alpha$ photons are trapped in this zone, there is a significant population of the $n = 2$ level of neutral hydrogen, and higher-n states are populated by collisions of electrons with neutral H in excited states; this gives rise to enhanced Balmer emission, and apparently explains the anomalously low values (relative to other photoionized gases) of the Lyα/Hα flux ratio observed in broad-line spectra.

While most of the BLR emission seems to arise in optically thick clouds, as noted at the beginning of this section, there is also some evidence for an optically thin BLR component.

(1) In some Seyfert galaxies, there are differences between the profiles of lines that are expected to be strong in optically thick clouds and those which can be strong in optically thin clouds (Morris and Ward 1989).

(2) While the core of the broad Hβ emission line in the spectrum of the Seyfert 1 galaxy Mrk 590 varies with the continuum, the high-velocity wings remain constant, which suggests that the higher radial-velocity gas is optically thin (Ferland, Korista, and Peterson 1990).

(3) The behavior of certain broad-line flux ratios with the continuum level is fairly easily reproduced by BLR models that contain an optically thin component (Shields, Ferland, and Peterson 1995).

An important caveat to keep in mind is that the fiducial parameters that have been deduced for BLR clouds are probably arrived at through strong selection effects. For example, the cloud properties that we infer for the C IV-emitting clouds are likely to be those of clouds that are optimal for production of C IV emission, even though clouds with a broad range of properties might exist in the BLR. Recent calculations by Baldwin *et al.* (1995) strongly suggest that this indeed may be the case.

5.4   Broad-Line Profiles

5.4.1   Logarithmic Profiles

In principle, the emission-line profiles should provide an important clue to the dynamics of the broad-line region, if we suppose that indeed the principal broadening mechanism is due to Doppler motions of individual clouds. To first order, the broad emission-line profiles in AGN spectra are logarithmic, i.e., the flux at a wavelength shift $\Delta\lambda$ from line center is proportional to $\log \Delta\lambda$. Following Capriotti, Foltz, and Byard (1980), we will illustrate how such profiles can arise by calculating the expected emission-line profile for an ensemble of optically thick (in the ionizing continuum) broad-line clouds which are undergoing steady-state radiatively driven outflow. We emphasize that this is only one of several ways that logarithmic profiles can be produced, and the calculation is meant to be illustrative.

We begin by assuming that the emission-line profile produced by each cloud is determined by thermal broadening (eq. 5.1). This is much narrower than the spectral resolution generally used in AGN studies (typically hundreds of kilometers per second), and therefore we can model the intrinsic line profile from each cloud as a δ-function. The emission-line profile for an ensemble of clouds is thus described by

$$L(\lambda) = 2\pi \int_{r_{\min}}^{r_{\max}} \int_{-1}^{1} n_{\mathrm{c}}\, j_{\mathrm{c}}\, r^2\, \delta\left[\lambda - \lambda_0 \left(1 + \frac{v\mu}{c}\right)\right] d\mu\, dr \tag{5.20}$$

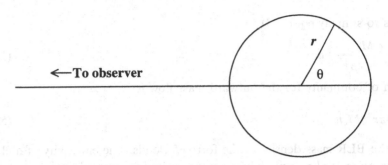

Fig. 5.4. Broad-line region clouds are located at distance r from the central source, at an angle θ measured from our line of sight to the central source.

where n_c is the number density of clouds in the BLR, each with emissivity j_c. The quantity λ_0 is the wavelength at line center, r_{min} and r_{max} are the inner and outer radii of the BLR, respectively ($v(r_{min})$ and $v(r_{max})$ will be the radial velocities of the clouds at these points), and $\mu = \cos\theta$, where θ is the angle between a BLR cloud and the point on the shell farthest away from the observer as measured from the center of the BLR (see Fig. 5.4); thus, the observer measures a Doppler shift (relative to the center of the BLR) of $\Delta\lambda = \lambda_0 v(r)\mu/c$ for a BLR cloud at position (r, θ).

The outward force on each cloud is given by

$$M_c\, a_{rad} = A_c \int_{v_1}^{\infty} \frac{L_v\, dv}{4\pi r^2 c} = \frac{A_c\, L_{UV}}{4\pi r^2 c}, \tag{5.21}$$

where M_c and A_c are the mass and cross-section for an individual cloud, and the integral is over the ionizing continuum, whose momentum is absorbed by the cloud. The radiative acceleration can also be written as

$$a_{rad} = \frac{dr}{dt}\frac{dv}{dr} = v\frac{dv}{dr}, \tag{5.22}$$

so we can write

$$dr = \frac{v\, dv}{a_{rad}} = \frac{4\pi r^2 c\, M_c}{A_c\, L_{UV}}\, v\, dv. \tag{5.23}$$

We can eliminate the quantity $A_c\, L_{UV}$ from this equation by considering the equation for photoionization equilibrium in an individual cloud. The number of photoionizations per second in the cloud equals the number of ionizing photons reaching the cloud, i.e.,

$$\frac{A_c}{4\pi r^2} \int_{v_1}^{\infty} \frac{L_v\, dv}{hv} = \frac{A_c}{4\pi r^2}\, Q(H), \tag{5.24}$$

and this must equal the total recombinations per second, $n_e^2 \alpha_B V_c$, where V_c is the ionized volume of the cloud. We can then define a mean energy per photon

$$\langle hv \rangle = \frac{L_{UV}}{Q(H)} = \frac{L_{UV}\, A_c}{4\pi r^2\, n_e^2 \alpha_B V_c}, \tag{5.25}$$

which allows us to simplify eq. (5.23) to

$$dr = \frac{c\,M_c}{\langle h\nu \rangle\, n_e^2 \alpha_B V_c}\, v\, dv. \tag{5.26}$$

The equation of continuity for the outward mass flow is

$$\frac{dM}{dt} = 4\pi r^2 v M_c n_c, \tag{5.27}$$

where $M_c n_c$ is the BLR mass density in the form of clouds. The emissivity of a cloud is given by integrating the line emissivity over the ionized volume of a cloud, i.e.,

$$j_c = \int\int j_{\text{line}}\, d\Omega\, dV = h\nu_{\text{line}}\, n_e^2 \alpha^{\text{eff}}\, V_c, \tag{5.28}$$

where ν_{line} is the frequency of the emitted line and α^{eff} is the effective recombination coefficient for the line.

We can now combine eqs. (5.26), (5.27), and (5.28) to obtain

$$
\begin{aligned}
n_c\, j_c\, r^2\, dr &= \left[\frac{dM/dt}{4\pi r^2 v\, M_c} \right] \left(h\nu_{\text{line}}\, \alpha^{\text{eff}}\, n_e^2 V_c \right) r^2 \left[\frac{c\,M_c}{\langle h\nu \rangle\, n_e^2 \alpha_B V_c}\, v\, dv \right] \\
&= \frac{h\nu_{\text{line}}}{\langle h\nu \rangle} \frac{\alpha^{\text{eff}}}{\alpha_B} \frac{c}{4\pi} \frac{dM}{dt}\, dv \\
&= K\, dv, \tag{5.29}
\end{aligned}
$$

where K is a constant. Inserting this in eq. (5.20) yields

$$
\begin{aligned}
L(\lambda) &= 2\pi K \int_{v(r_{\min})}^{v(r_{\max})} \int_{-1}^{1} \delta\left[\lambda - \lambda_0 \left(1 + \frac{v\mu}{c} \right) \right] d\mu\, dv \\
&= \frac{2\pi c}{\lambda_0} K \int_{v(r_{\min})}^{v(r_{\max})} \int_{-v\lambda_0/c}^{v\lambda_0/c} \delta\left[(\lambda - \lambda_0) - x \right] \frac{dx\, dv}{v} \\
&= \frac{2\pi c}{\lambda_0} K \int_{a}^{v(r_{\max})} \frac{dv}{v}. \tag{5.30}
\end{aligned}
$$

The integral is non-zero only when the condition $\lambda - \lambda_0 = \lambda_0 v\mu/c$ is met, i.e., for values

$$\mu = \frac{(\lambda - \lambda_0)c}{\lambda_0 v}. \tag{5.31}$$

If we integrate over the range $-1 \leq \mu \leq 1$, there are two values of μ where the integrand is non-zero, at $\pm(\lambda - \lambda_0)c/v\lambda_0$, i.e., $|\lambda - \lambda_0|\, c/v\lambda_0$. However, the lower limit on the integral over v must now be changed in that clouds at velocities smaller than $c\,|\lambda - \lambda_0|\,/\lambda_0$ do not contribute to the emission line profile at λ; thus the lower limit becomes $a = v(r_{\min})$ or $|\lambda - \lambda_0|\, c/\lambda_0$, whichever is larger. Taking $v(r_{\min}) = 0$ gives

$$L(\lambda) = \frac{2\pi c}{\lambda_0} K\, \ln\left(\frac{v(r_{\max})\lambda_0}{|\lambda - \lambda_0|\, c} \right), \tag{5.32}$$

i.e., a logarithmic profile.

Unfortunately, the line profiles alone tell us very little about the BLR velocity field, as there are a number of different scenarios, mostly involving radial motion (but in either direction), which can produce logarithmic profiles (Capriotti, Foltz, and Byard 1980). The line profiles alone do not provide a strong discriminant, only a constraint that must be met by any model.

5.4.2 Line Asymmetries and Wavelength Shifts

The general logarithmic shape of the line profiles is also an oversimplification, in the same way that it is an oversimplification to describe the spectral energy distribution as a power law. Real line profiles often show asymmetries, which can be very pronounced. For example, a redward asymmetric profile (more line emission redward of line center than blueward) might be accounted for by either (a) inflow of clouds in an attenuating medium, or (b) outflow of clouds in which most of the line emission comes from the inward-facing side of the BLR clouds. Unfortunately, the situation is very complex as real line profiles often show both blueward and redward asymmetries (i.e., different asymmetries are seen in different sources) and structure that suggests that the BLR might be comprised of a number of distinct physical components. Moreover, the line profiles can vary with time, and profiles in a single galaxy can change the sense of their asymmetry over time. There was early speculation that this might be due to excitation inhomogeneities in the BLR on account of continuum variability, but it is now clear that the most pronounced profile changes occur on time scales longer than the light-travel time across the BLR. The nature of profile variability is not yet understood.

The profiles of various lines in an AGN spectrum can differ significantly from one another in detail. For example, in general He II $\lambda 4686$ is significantly broader than the Balmer lines (by as much as 50%). Even the Balmer lines have different profiles, with Hβ broader than Hα; another way of looking at this is that the Balmer decrement varies with line-of-sight velocity, such that Hβ/Hα increases with Δv from values typically ~ 5 near line center to less than ~ 2 in the line wings (Crenshaw 1986, Stirpe 1991). A potentially important observation is that the peaks of various emission lines are observed to occur at different radial velocities, i.e., the redshifts measured from various emission lines are not the same (Gaskell 1982, Wilkes 1984, Corbin 1990, Tytler and Fan 1992). The general sense of these wavelength shifts is that the higher-ionization lines (e.g., C IV $\lambda 1549$) are shifted blueward relative to the lower-ionization lines (e.g., Mg II $\lambda 2798$), which are in turn at approximately the *systemic* redshift of the AGN, as measured by the forbidden lines and/or the absorption lines in the host galaxy (see Chapter 6). The magnitude of these wavelength shifts is often several hundred kilometers per second, with some shifts as large as $10^3 \, \mathrm{km \, s^{-1}}$ or more. The existence of such discrepancies seems to imply (a) that the line profiles are affected by some combination of radial motion and line attenuation, and (b) that the radial motion and attenuation are not the same for every emission line, which suggests a stratified rather than homogeneous BLR structure. The latter conclusion is supported by the results of line-reverberation studies, as discussed in the next section.

5.5   Reverberation Mapping

In principle, the size and structure of the BLR can be obtained by observing the response of emission lines to continuum variations. The emission-line response to a continuum change is delayed and spread out in time by light travel-time effects within the BLR – the BLR thus appears to 'reverberate' in response to the continuum variations.

Consider as an example the response of a thin-shell BLR as shown in Fig. 5.4 to a sudden continuum outburst (i.e., a δ-function in time). The continuum pulse is then localized in a spherical shell that expands outwards from the central source at the speed of light. After a time $t = r/c$ the continuum pulse will reach the BLR clouds, which absorb the incident radiation and reprocess it into emission-line photons on a time scale given by the hydrogen recombination time,

$$\tau_{\text{rec}} = (n_e \alpha_B)^{-1} \approx 40 \, n_{11}^{-1} \text{ sec}, \tag{5.33}$$

where the electron density has been expressed as $n_{11} = n_e/10^{11} \text{ cm}^{-3}$. Thus, the ionizing photons in the continuum burst are locally reprocessed into line photons virtually instantaneously. The continuum outburst will eventually be recorded by a distant observer along the $-x$ axis. The emission-line response from gas clouds on the $-x$ axis will be recorded simultaneously. The response from other parts of the BLR will arrive later on account of the longer total path from the continuum to the BLR clouds and then to the observer. At some fixed time delay τ after the arrival of the continuum outburst at the observer's location, all of the line photons reaching the observer will come from gas clouds that lie along a surface of constant time delay (or 'isodelay surface'), which must be a paraboloid whose axis coincides with the observer's line of sight to the continuum source. The BLR response at time delay τ is thus given by the intersection of the BLR gas distribution and the appropriate isodelay paraboloid

$$\tau = (1 + \cos \theta) r/c. \tag{5.34}$$

For a thin spherical shell of radius r, the intersection of these surfaces is a ring of radius $r \sin \theta$ with surface area $2\pi r^2 \sin \theta \, d\theta$, as shown in Fig. 5.5. If we assume that the emission-line 'responsivity' (which we can think of as the number of extra photons produced in a given emission line for a given increase in the continuum level) of all of the clouds is the same, the responsivity per unit area of the BLR shell ζ is constant. The emission-line response seen by the observer is the product of ζ and the area of the ring. The emission-line response as a function of θ is

$$\Psi(\theta) \, d\theta = 2\pi \zeta r^2 \sin \theta \, d\theta. \tag{5.35}$$

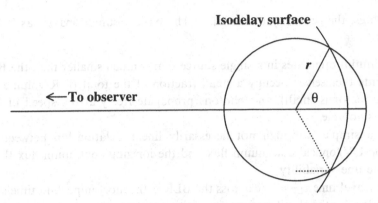

Fig. 5.5. In this idealized model, the BLR clouds are assumed to be located in a thin spherical shell of radius r centered on the continuum source. As seen by a distant observer, the response of the BLR clouds to a change in the continuum brightness is delayed by light travel-time effects. Continuum photons traveling radially outward along the dotted line reach the BLR and are instantaneously reprocessed into emission-line photons, some of which are emitted in the observer's direction, as shown by the continuation of the dotted line. The observed line response is thus delayed by the light-travel time along the dotted path, i.e., $\tau = (1 + \cos\theta)r/c$. At some fixed τ, the observer sees the line response from all clouds that lie along the 'isodelay surface' given by this equation, which is for a paraboloidal surface.

From eq. (5.34), the range of values of θ which corresponds to fixed infinitesimal units of time is given by

$$d\tau = -(r/c)\sin\theta\, d\theta. \tag{5.36}$$

The line response can then be written as a function of time delay as

$$\Psi(\tau)\, d\tau = \Psi(\theta) \left| \frac{d\theta}{d\tau} \right| d\tau = 2\pi\zeta rc\, d\tau, \tag{5.37}$$

where the emission-line response at time τ is given by $\Psi(\tau)$, which is called the 'transfer function'. Thus, the transfer function for a thin spherical shell is particularly simple; the line response is constant between $\tau = 0$ (i.e., $\theta = 180°$), corresponding to the clouds along the line of sight to the observer and $\tau = 2r/c$ (i.e., $\theta = 0°$), corresponding to the far side of the BLR as seen by the observer. The reason for this is that while the ring which is the locus of intersection has its largest radius around $\theta = 90°$, by eq. (5.36), a fixed interval of time $d\tau$ corresponds to the smallest range in $d\theta$ at the same time. The $\sin\theta$ factor cancels out of the equation when the response is expressed in units of time delay. Transfer functions for other geometries are given by Blandford and McKee (1982).

The reverberation-mapping technique does not depend on any particular BLR geometry. Indeed, the goal of reverberation mapping is to determine the transfer function from the emission-line response to the continuum variability and thus infer, among other things, the geometry of the BLR. The basic assumptions are as follows (Peterson 1993):

(1) The continuum originates in a single source that is much smaller than the BLR.
(2) BLR clouds themselves occupy a small fraction of the total BLR volume (i.e., the filling factor is small), and photons propagate freely at the speed of light within this volume.
(3) There is a simple, although not necessarily linear, relationship between the observable UV/optical continuum flux and the ionizing continuum flux that is driving the line variability.
(4) The light-travel time $\tau_{LT} = r/c$ across the BLR is the most important time scale. In particular, (a) the cloud response to continuum variations is short compared to τ_{LT} and (b) τ_{LT} is short compared to the time scale on which significant geometrical changes in the BLR can occur (i.e., the dynamical time scale for the BLR). The time scale for a cloud to reprocess the ionizing radiation into line photons is given by τ_{rec} (eq. 5.33), which is virtually instantaneous. The dynamical time scale can be approximated from the time it would take a line-emitting cloud to cross the BLR, $\tau_{dyn} \approx r/\Delta v_{FWHM}$, so $\tau_{dyn}/\tau_{LT} \approx c/\Delta v_{FWHM} \approx 100$.

In the real case, the observer sees at some fixed time t line radiation from all isodelay surfaces, with the response of each surface a function of the continuum level at a different time in the past. The emission-line flux at time t is thus given by integrating over all of the isodelay surfaces, i.e.,

$$L(t) = \int_{-\infty}^{\infty} \Psi(\tau)C(t - \tau)\, d\tau, \tag{5.38}$$

which is known as the 'transfer equation'. Obviously the transfer function is the emission-line response to a δ-function continuum pulse. The goal of reverberation mapping of the BLR is to use the observables, the continuum light curve $C(t)$ and the emission-line light curve $L(t)$, to solve for the transfer function $\Psi(\tau)$. A large amount of high-quality data is necessary to obtain a unique solution to an integral equation like eq. (5.38), and even the best existing data yield results which are somewhat ambiguous. However, with relatively few or lesser quality data it is nevertheless possible to obtain some measure of the spatial extent of the BLR by simply cross-correlating the continuum and emission-line light curves in order to find the temporal shift between them that maximizes the correlation. This shift is usually referred to as the 'lag' for the particular emission line relative to the continuum. The cross-correlation function is formally defined as

$$F_{CCF}(\tau) = \int_{-\infty}^{\infty} L(t)C(t - \tau)\, dt. \tag{5.39}$$

The continuum autocorrelation function is defined similarly as

$$F_{ACF}(\tau) = \int_{-\infty}^{\infty} C(t)C(t - \tau)\, dt. \tag{5.40}$$

Table 5.2

NGC 5548

Cross-Correlation Lags

Feature	τ (Days)
N v λ1240	2
He II λ1640	2
'Small blue bump'	6
He II λ4686	7
He I λ5876	9
Lyα λ1216	10
C IV λ1549	10
Hγ λ4340	13
Hα λ6563	17
Hβ λ4861	20
C III] λ1909	22

By using eq. (5.38) for $L(t)$ in eq. (5.39), we get

$$F_{\rm CCF}(\tau) = \int_{-\infty}^{\infty} C(t-\tau) \int_{-\infty}^{\infty} \Psi(\tau')C(t-\tau')d\tau'dt. \qquad (5.41)$$

By reversing the order of integration, we obtain

$$F_{\rm CCF}(\tau) = \int_{-\infty}^{\infty} \Psi(\tau') \int_{-\infty}^{\infty} C(t-\tau')C(t-\tau)dtd\tau', \qquad (5.42)$$

and the interior integral is now recognized as eq. (5.40) for $F_{\rm ACF}(\tau - \tau')$, so that

$$F_{\rm CCF}(\tau) = \int_{-\infty}^{\infty} \Psi(\tau')F_{\rm ACF}(\tau - \tau')\,d\tau'. \qquad (5.43)$$

Thus, the cross-correlation function is the convolution of the transfer function and the continuum autocorrelation function. The lag for a particular emission line is defined to be the location of either the peak or centroid of the cross-correlation function $F_{\rm CCF}(\tau)$. The relationship between the cross-correlation function and the transfer function is thus fairly simple, although it is still clear upon some reflection that the lag measured depends not only on the transfer function, but also on the characteristics of the continuum variability, as characterized by $F_{\rm ACF}(\tau)$. By measuring the cross-correlation lag, we can obtain some basic information about the transfer function, namely a scale length for the BLR based on the mean response time for a given emission line.

One of the most important results of reverberation-mapping experiments is the demonstration that the responsivity distributions are different for different lines (see Fig. 5.6 and Table 5.2), i.e., different lines have different lags. Lines that are prominent in highly ionized gases (e.g., He II λ1640, N v λ1240, C IV λ1549) respond to continuum

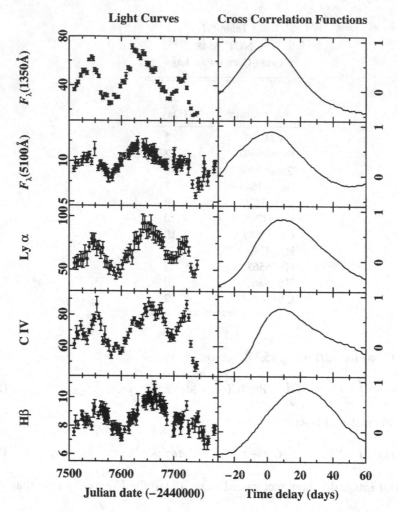

Fig. 5.6. The left-hand column shows continuum and emission-line light curves for the Seyfert 1 galaxy NGC 5548 that were obtained in 1988–89 (Clavel *et al.* 1991, Peterson *et al.* 1991). The ultraviolet (1350 Å) and optical (5100 Å) fluxes (in units of 10^{-15} ergs s^{-1} cm^{-2} Å^{-1}) appear to vary in phase. The variations of the strong emission lines (Lyα λ1216, C IV λ1549, and Hβ, in units of 10^{-13} ergs s^{-1} cm^{-2}) show the same general pattern of variations, but with a time delay due to light travel-time effects in the BLR. The time delay is calculated by cross-correlation of each light curve with the 1350 Å light curve. The cross-correlation functions are shown in the right-hand column for each light curve. The panel to the right of the 1350 Å shows the autocorrelation function, i.e., the result of cross-correlating the 1350 Å continuum with itself.

variations faster than lines that are most prominent at lower ionization levels (e.g., the Balmer lines). This clearly indicates that ionization structure of the BLR is radially stratified. In particular, as pointed out earlier the responsivity distributions of C III] and C IV are different and these lines arise in different parts of the BLR.

To a low order of approximation, AGN broad-line spectra are quite similar over several orders of magnitude in luminosity, from low-luminosity Seyfert 1s to high-luminosity QSOs (see, however, §5.8.2), which then suggests that U and n_e are about the same in all BLRs. If this is true, eqs. (5.14) and (5.15) then lead to the naïve prediction that

$$r \propto Q^{1/2} \propto L^{1/2}. \tag{5.44}$$

Normalizing this equation to C IV in NGC 5548 gives

$$r = 10h_0 \left[\frac{L_\lambda(1350\,\text{Å})}{10^{40}\,\text{ergs s}^{-1}\,\text{Å}^{-1}} \right]^{1/2} \text{light days}, \tag{5.45}$$

or, ignoring the C IV luminosity dependence (§5.8.2), we obtain eq. (5.6). Cross-correlation of continuum and emission-line light curves for a number of different AGNs seem at least consistent with this prediction, although the well-observed sources are few and the range of luminosity that has been explored is limited.

5.6   The BLR Velocity Field

With fairly accurate BLR sizes known from reverberation mapping, we could in principle determine a virial mass for the central source (eq. 1.1) if we also knew the BLR velocity field (i.e., $v = f(r, \theta, \phi)$). The gross velocity field of the BLR could be either radial outflow or infall, or the cloud motions might be primarily orbital, either ordered (e.g., in a disk) or random. At least an order-of-magnitude mass estimate could be made as long as radial outflow could be excluded, so that presumably the dominant force acting on the clouds is gravity, not, for example, radiation pressure.

As noted in §5.4, the line profiles are not uniquely characteristic of any particular velocity field, so they contain very little information. Ideally, we would like to generalize the reverberation-mapping formalism to include velocity dependence and then solve

$$L(v, t) = \int_{-\infty}^{\infty} \Psi(v, \tau) C(t - \tau)\, d\tau. \tag{5.46}$$

As in the case of eq. (5.38), this has yet been done convincingly. Alternatively, the broad emission lines can be divided up into several different radial-velocity ranges and the resulting light curves can be cross-correlated with the continuum light curve or relative to one another so that the emission-line lag can be measured as a function of radial velocity. A few attempts (e.g., Korista *et al.* 1995, Wanders *et al.* 1995) along these lines seem to indicate that *predominantly* radial motions of the C IV-emitting clouds can be excluded. It is thus probably safe to estimate the mass of the central object using the results for the C IV emission line in NGC 5548 ($r \approx 10$ light days, $\Delta v_{\text{FWHM}} \approx 4500\,\text{km s}^{-1}$),

$$M \approx \frac{rv^2}{G} \approx 8 \times 10^{40}\,\text{g} \approx 4 \times 10^7\,M_\odot. \tag{5.47}$$

We might also assume that changes in the emission-line profiles might reveal something about the net BLR velocity field, and indeed this provided much of the early motivation for emission-line variability studies. However, as large amounts of data have become available on a few bright AGNs, it has become apparent that the line-profile changes are not reverberation effects, and they are not correlated with the continuum variability in any obvious way.

5.7   Cloud Properties

Thus far we have left unaddressed the fundamental problem of the nature of the BLR clouds. It is therefore useful at this point to consider what information we might have that could lead to estimates of their sizes, numbers, and physical conditions. As a specific example, we will carry out some very approximate calculations in which we will assume a C IV luminosity of 10^{42} ergs s^{-1} and a BLR size $r \approx 8$ light days. We will also ignore any dependence on h_0.

First, our calculation of the BLR covering factor allows a constraint on the cloud sizes and number through eq. (5.4). If we assume that the BLR clouds are arranged in a spherical shell surrounding the central source at a distance r and that the cross-section for an individual cloud is $\pi \ell^2$, then the fraction of the sky covered by BLR clouds (assuming that they do not shadow one another) is

$$f \approx \frac{N_c \pi \ell^2}{4\pi r^2} \approx \frac{N_c \ell^2}{1.6 \times 10^{33}} \approx 0.1, \tag{5.48}$$

or

$$N_c \left(\frac{\ell}{R_\odot} \right)^2 \approx 3 \times 10^{10}. \tag{5.49}$$

An independent estimate can be made by comparing the total luminosity in C IV with that of a single cloud (cf. eq. 5.3)

$$\begin{aligned} L_{\text{cloud}}(\text{C IV}) &= V \, n_e n_{\text{C}^{3+}} \, q(2s\,{}^2\text{S}, 2p\,{}^2\text{P}) h\nu \\ &\approx V \, n_e^2 \, 1.1 \times 10^{-23} \text{ ergs s}^{-1}, \end{aligned} \tag{5.50}$$

where V is the volume of a single cloud. We take the emitting volume to be an ionized skin of depth r_1 (eq. 5.18) on a large cloud,† i.e., $V = 2\pi \ell^2 r_1$. With our fiducial density of $n_e = 10^{11}$ cm^{-3}, we obtain

$$L_{\text{cloud}}(\text{C IV}) \approx 3.7 \times 10^{10} \, \ell^2 \text{ ergs s}^{-1}, \tag{5.51}$$

where we have used eq. (5.19) for r_1. The number of clouds is given by the total C IV

---

† In some treatments, the emitting volume is simply assumed to go like ℓ^3, in which case the calculations given here will lead to a condition on the product $N_c \ell^3$. By comparing this with eq. (5.49), one infers a very large number ($\gtrsim 10^{11}$) of small clouds (see Peterson 1994).

luminosity (10^{42} ergs s^{-1}) divided by the luminosity per cloud, which yields

$$N_c \left(\frac{\ell}{R_\odot} \right)^2 \approx 6 \times 10^9. \tag{5.52}$$

Given the approximate nature of these calculations, eqs. (5.49) and (5.52) are in excellent agreement.

Finally, we also point out that the number of clouds in the BLR can be estimated very crudely by noting that the line profiles are often very smooth and show little if any structure at high resolution (even the structures seen in some profiles as in Fig. 5.1 are quite smooth and extend over several resolution elements). This means that the statistical fluctuations in the number of clouds per resolution element must be smaller than the signal-to-noise ratio. Simulations of observations (Atwood, Baldwin, and Carswell 1982, Capriotti, Foltz, and Byard 1981), based on the assumption that individual clouds have line profiles with thermal widths (eq. 5.1), suggest that the number of clouds must be at least $N_c \approx 5 \times 10^4$, which combined with eqs. (5.49) or (5.52) imply $\ell \lesssim 400\, R_\odot$.

What can be said about the nature of the BLR clouds? The prevailing view in the early 1980s was that the BLR clouds represent dense condensations that are in pressure equilibrium (i.e., $n_e T$ constant) with a hotter low-density intercloud medium that serves to confine the BLR clouds (Krolik, McKee, and Tarter 1981). Some method of confining the clouds is necessary, since they are much too small to be self-gravitating; the Jeans mass† for a 10^4 K gas at BLR densities is more than $\sim 100\, M_\odot$, which is much greater than the line-emitting mass of the entire BLR. This 'two-phase' model is no longer widely accepted as plausible primarily on account of the high temperature required for the intercloud medium (e.g., Mathews and Ferland 1987). Magnetic confinement of clouds has also been considered (e.g., Rees 1987). In any case, some kind of confinement or replenishment mechanism must be at work since the clouds are otherwise subject to a variety of instabilities (including thermal evaporation) that can destroy them on time scales shorter than τ_{dyn} (Mathews and Capriotti 1985).

An interesting alternative is to consider the possibility that the BLR clouds are associated with stars, with the broad-line emission originating in extended envelopes or perhaps in material ablated off the stars (e.g., Penston 1988, Kazanas 1989, and references therein). The major problem that has been encountered so far is that the process seems to work effectively only for low surface-gravity giant stars; if these constitute only 10^{-4} of the mass of the population and the number of clouds required is fairly large, the mass of stars within the BLR becomes uncomfortably large. For example, if $\ell \approx 100\, R_\odot$, then $N_c \approx 10^6$; if each of these clouds represents a giant star in a typical stellar population, $M_{\mathrm{BLR}} \approx 10^{10}\, M_\odot$.

---

† The Jeans mass for an ionized cloud of hydrogen is

$$M_J = \frac{\pi^{5/2}}{6} \left(\frac{kT}{Gm_p} \right)^{3/2} \rho^{-1/2},$$

where ρ is the mass density. This is the minimum mass that can collapse under self-gravity.

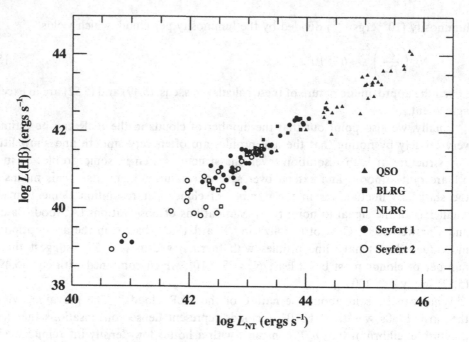

Fig. 5.7. The total Hβ line luminosity is proportional to the AGN continuum luminosity (here called the 'non-thermal continuum luminosity' L_{NT}) over the entire range of AGN luminosities. Values of $h_0 = 1$ and $q_0 = 0.5$ (see Chapter 9) are assumed. Adapted from Yee (1980).

5.8   Line–Continuum Correlations

The gross similarity of BLR spectra over a wide range of luminosity obviously implies that the emission-line luminosities are quite strongly correlated with the continuum luminosities; another way of saying this is that the emission-line equivalent widths are approximately the same from object to object.

5.8.1   The Balmer Lines

The correlation between the Hβ luminosity and the continuum luminosity has been examined by several authors. Yee (1980) has investigated this relationship and finds that the 'non-thermal' continuum (i.e., after removal of the starlight) over the range ~ 3000–$9500\,\text{Å}$ has luminosity

$$L_{NT} \approx 80\,L(H\beta). \tag{5.53}$$

This relationship, which is shown in Fig. 5.7, provided one of the original very strong arguments that the BLR is photoionized by the central source. Figure 5.7 clearly shows the continuity between Seyferts and quasars, and the overlap in luminosity between them.

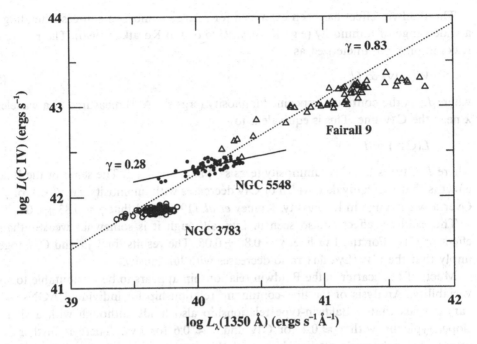

Fig. 5.8. The Baldwin effect in C IV λ1549. The correlation between the C IV emission-line and 1350 Å continuum luminosities for three variable Seyfert galaxies, NGC 3783 (Reichert *et al.* 1994), NGC 5548 (Clavel *et al.* 1991), and Fairall 9 (Clavel, Wamsteker, and Glass 1989). The individual points correspond to repeated observations over many months with *IUE*. Different symbols are used for each of these AGNs. Over a wide range in mean continuum luminosity, the C IV luminosity does not scale linearly with the continuum luminosity; this is the Baldwin effect. The dotted line shows the slope of the Baldwin effect ($\gamma = 0.83$, see eq. 5.55) as determined by Kinney, Rivolo, and Koratkar (1990). Also shown in the shallower slope ($\gamma = 0.28$, from Pogge and Peterson 1992) of a similar relationship that is seen as a single source varies in luminosity. Much of the scatter around the latter relationship is due to light travel-time effects in the broad-line region. Values of $h_0 = 1$ and $q_0 = 0.5$ (see Chapter 9) are assumed.

5.8.2   The Baldwin Effect

Baldwin (1977) showed that the equivalent width (eq. 1.10) of the C IV emission line tends to decrease systematically with increasing continuum luminosity. This inverse correlation between the C IV equivalent width and luminosity is known as the 'Baldwin effect'. The origin of this effect is not understood. Suggested explanations include (a) decrease in U (ionization parameter) with luminosity, (b) decrease in f (covering factor) with luminosity, and (c) disk-inclination effects; in the last case, the continuum is assumed to arise in an accretion disk which appears brighter when seen at lower inclination, but the emission-line flux comes from an extended region which radiates isotropically.

The Baldwin effect has been examined for a large number of sources extending over a wide range of luminosity (e.g., Kinney, Rivolo, and Koratkar 1990). The relationship is commonly parameterized as

$$W(\text{C\,IV}) \propto L_\lambda^\beta, \tag{5.54}$$

where L_λ is the continuum specific luminosity (ergs s^{-1} Å^{-1}) measured at a wavelength λ near the C\,IV line. This is equivalent to

$$L(\text{C\,IV}) \propto L_\lambda^\gamma, \tag{5.55}$$

where $L(\text{C\,IV})$ is the C\,IV luminosity (ergs s^{-1}) and $\gamma = \beta + 1$. The sense of the Baldwin effect is that the equivalent width of C\,IV decreases with luminosity, i.e., $\beta < 0$ or $\gamma < 1$. Over a wide range in luminosity, Kinney *et al.* (1990) find that $\gamma \approx 0.83 \pm 0.04$.

The Baldwin effect is also seen in Lyα, although it is somewhat weaker than the effect in C\,IV. For the Lyα line, $\gamma \approx 0.88 \pm 0.05$. The results for Ly$\alpha$ and C\,IV together imply that the C\,IV/Lyα flux ratio decreases with luminosity.

Much of the scatter in the Baldwin relationship appears to be attributable to source variability. Analysis of the line–continuum relationship for individual AGNs as they vary reveals that a Baldwin-type relationship also holds, although with a shallower slope, typically with $\gamma \approx 0.4$ for C\,IV and $\gamma \approx 0.6$ for Lyα. There is further scatter around these *intrinsic* relationships, and this appears to be due to light travel-time effects within the BLR; when the line luminosities at time t are compared with the continuum luminosities at time $t - \tau$, where τ is the reverberation lag, the relationship has almost no scatter (Pogge and Peterson 1992). The shallower slope of this intrinsic relationship introduces scatter in the object-to-object relationship. The Baldwin effect is illustrated in Fig. 5.8.

From an observational point of view, the importance of the Baldwin effect is its potential as a 'standard candle' in cosmological investigations. If the equivalent width of an emission line allows one to infer the source luminosity, a plot of luminosity versus flux will be sensitive to cosmological parameters such as q_0 (Chapter 9), thus providing an important consistency check for cosmological models.

6   The Narrow-Line Region

The narrow-line region (NLR) in AGNs is of interest for at least three interrelated reasons. First, the NLR is the largest spatial scale where the ionizing radiation from the central source dominates over other sources. Second, the NLR is the only AGN component which is spatially resolved in the optical – this is of particular importance as the NLR is clearly illuminated in a non-isotropic manner by the central source. Finally, the NLR dynamics might tell us something about how AGNs are fueled.

6.1   Narrow-Line Spectra

As in the case of the BLR, the relative strengths of the emission lines we observe in NLR spectra allow us to discern some of the properties of the ionizing spectrum. Unlike the BLR, the electron densities in the NLR are low enough that many forbidden transitions are *not* collisionally suppressed. This allows us to use the intensity ratios of certain pairs of forbidden lines to measure the electron densities and temperatures in the NLR gas (§6.2). In comparison to the BLR, the analysis is simplified by the low densities. On the other hand, however, in the case of the NLR an additional complication is introduced into the spectroscopic analysis by significant amounts of dust since the NLR arises outside the dust sublimation radius (eq. 4.15); indeed, it may well be that the radius where dust sublimates provides the fundamental demarcation between the BLR and the NLR.

General properties of the NLR emission that should be kept in mind are (a) that the emission comes from a spatially extended region, so that at least to some extent physical and kinematic distributions can be mapped out directly and (b) that the forbidden-line emission is isotropic since self-absorption in narrow lines is negligible. Only dust on large spatial scales can result in orientation-dependent effects on NLR line fluxes.

Table 6.1 contains a list of emission lines that are generally prominent in narrow-line spectra, along with their typical relative intensities in Seyfert 2 spectra (from Ferland and Osterbrock 1986). Also given in Table 6.1 is the relevant ionization potential for each line. In the case of collisionally excited lines, the ionization potential is that necessary to achieve the observed ionization state. For recombination lines, the relevant ionization potential is that needed to achieve the *next higher* state of ionization, since it is recombination from these states that leads to the formation of the emission lines.

Table 6.1

Strong Narrow Lines in Seyfert Spectra

Line	Relative Flux	Ionization Potential (eV)	Critical Density n_{crit} (cm^{-3})
Lyα λ1216	55	13.6	...
C IV λ1549	12	47.9	...
C III] λ1909	5.5	24.4	3.0×10^{10}
Mg II λ2798	1.8	7.6	...
[Ne V] λ3426	1.2	97.1	1.6×10^{7}
[O II] λ3727	3.2	13.6	4.5×10^{3}
[Ne III] λ3869	1.4	40.0	9.7×10^{6}
[O III] λ4363	0.21	35.1	3.3×10^{7}
He II λ4686	0.29	54.4	...
Hβ λ4861	1.00	13.6	...
[O III] λ4959	3.6	35.1	7.0×10^{5}
[O III] λ5007	11	35.1	7.0×10^{5}
[N I] λ5199	0.15	0.0	2.0×10^{3}
He I λ5876	0.13	24.6	...
[Fe VII] λ6087	0.10	100	3.6×10^{7}
[O I] λ6300	0.57	0.0	1.8×10^{6}
[Fe X] λ6375	0.04	235	4.8×10^{9}
[N II] λ6548	0.9	14.5	8.7×10^{4}
Hα λ6563	3.1	13.6	...
[N II] λ6583	2.9	14.5	8.7×10^{4}
[S II] λ6716	1.5	10.4	1.5×10^{3}
[S II] λ6731	...	10.4	3.9×10^{3}
[Ar III] λ7136	0.24	27.6	4.8×10^{6}

A wide variety of ionization states are present in narrow-line spectra. Both low-ionization lines (e.g., [O I] λ6300) and high-ionization lines (e.g., [O III] $\lambda\lambda$4959, 5007) are strong, and indeed in some cases very highly ionized species are detected (e.g., [Fe VII] λ6087, [Fe X] λ6375, [Fe XI] λ7892, and [Fe XIV] λ5303, the iron 'coronal' lines†). The [O III] λ5007/Hβ flux ratio is usually larger than 3. These simple observations, and the unique location of the narrow-line flux ratios of AGNs on BPT diagrams (e.g., Fig. 2.3) indicate that the NLR gas is indeed photoionized by the AGN spectrum, and not by stars in the nuclear regions.

The full width at half maximum for narrow emission lines falls in the range $200 \lesssim \Delta v_{FWHM} \lesssim 900 \, \text{km s}^{-1}$, with most values falling around 350–400 km s^{-1}. The

---

† These lines all have critical densities (§6.2.1) in the range 10^7–10^{10} cm^{-3}, and were first detected in the solar corona, which explains their name.

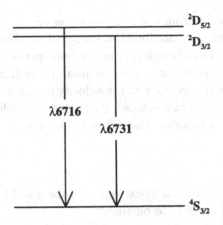

Fig. 6.1. A partial energy-level diagram for S^+ showing the transitions that produce the [S II] $\lambda\lambda6716$, 6731 doublet. Adapted from Osterbrock (1989), p. 132.

well-known and almost prototypical Seyfert 2 galaxy NGC 1068 stands out as extreme in this distribution with $\Delta v_{\text{FWHM}} \approx 1200 \, \text{km s}^{-1}$.

6.2 Physical Conditions in Low-Density Gases

In this section, we briefly review how electron densities and temperatures in ionized gases can be determined by the relative strength of various non-electric-dipole radiative transitions out of collisionally excited metastable levels. A detailed discussion can be found in Osterbrock (1989, Chapter 5).

6.2.1 Electron Densities

Electron densities are determined by measuring the intensity ratio of two lines from a single ion (to obviate ambiguities due to chemical composition or ionization level) which arise in closely spaced upper states and decay to a common lower state. Well-known examples of such pairs are [O II] $\lambda\lambda3726$, 3729 and [S II] $\lambda\lambda6716$, 6731, which in each case represents $^4S_{3/2}-^2D_{5/2}$ and $^4S_{3/2}-^2D_{3/2}$ transitions, although the [O II] lines are so close together that in AGNs they are heavily blended by Doppler broadening and are therefore not of much practical use. We show the level structure of S^+ in Fig. 6.1.

The basic physics involved can be illustrated through consideration of a simple two-level atom with excitation potential χ. The emissivity in the line due to the downward transition $2 \rightarrow 1$ is

$$j_{21} = n_2 A_{21} \frac{h\nu_{21}}{4\pi} \quad \text{ergs s}^{-1} \text{ cm}^{-3} \text{ ster}^{-1}. \tag{6.1}$$

Here n_2 is the number density (cm^{-3}) of atoms in the $n = 2$ level, A_{21} is the Einstein

coefficient for a spontaneous radiative transition from $n = 2$ to $n = 1$ (per second), and $h\nu_{21}$ is the energy of a photon resulting from the transition. In statistical equilibrium the rate at which the $n = 2$ level is populated by collisions (per unit volume per second) is $q_{12} \equiv \langle \sigma_{12} v \rangle$, where σ_{12} is the (velocity-dependent) cross-section for collisional excitation of the $n = 2$ level and the average is over the electron velocity distribution.†
The collisional excitation rate is is balanced by the rate at which the level is depopulated by subsequent collisions and by spontaneous radiative transitions to the $n = 1$ level, i.e.,

$$\langle \sigma_{12} v \rangle n_e n_1 = n_2 A_{21} + \langle \sigma_{21} v \rangle n_e n_2, \tag{6.2}$$

where σ_{21} is obviously the cross-section for collisional de-excitation of the $n = 2$ level. We can solve eq. (6.2) for n_2 and use this in eq. (6.1) to obtain

$$j_{21} = n_e n_1 \langle \sigma_{12} v \rangle \frac{A_{21}}{A_{21} + n_e \langle \sigma_{21} v \rangle} \frac{h\nu_{21}}{4\pi}. \tag{6.3}$$

We also know from the principle of detailed balance that

$$\langle \sigma_{12} v \rangle = \langle \sigma_{21} v \rangle \frac{g_2}{g_1} e^{-\chi/kT_e}, \tag{6.4}$$

where g_n is the statistical weight of the level n ($g = 2J + 1$, where J is the total angular momentum quantum number), and T_e is the electron temperature. The exponential term on the right-hand side appears because the electrons must have a threshold kinetic energy χ to collisionally excite the $n = 2$ level.

We now consider two separate cases which depend on the relative size of the two terms in the denominator of eq. (6.3). The only variable in these terms is the electron density n_e since all the other parameters are physical constants, so we distinguish two regimes based only on the electron density:

Low-density case: If the electron density is low, then $n_e \langle \sigma_{21} v \rangle \ll A_{21}$, in which case eq. (6.3) simplifies to

$$j_{21} = n_e n_1 \langle \sigma_{12} v \rangle \frac{h\nu_{21}}{4\pi} = n_e n_1 q_{12} \frac{h\nu_{21}}{4\pi}. \tag{6.5}$$

In the low-density case, eq. (6.2) shows that the radiative de-excitation rate is much greater than the collisional de-excitation rate. In other words, the radiative process is comparatively so fast that virtually all collisional excitations lead to a radiative de-excitation. From eq. (6.5), we see that in the low-density case, $j \propto n^2$.

---

† The mean free path for an electron to collide with an ion is $\ell = 1/n_i \sigma$, where n_i is the number density of ions (cm^{-3}) and σ is their cross-section (cm^2). The mean time between collisions is thus $t = \ell/v = 1/n_i \sigma v$, so the rate at which collisions occur per unit volume is $n_e/t = n_e n_i \sigma v$ cm^{-3} s^{-1}. Since the cross-section for collisional excitation is a function of velocity, the mean rate per unit volume must be averaged over the electron velocity distribution, i.e.,

$$q(\text{cm}^3 \, \text{s}^{-1}) \equiv \langle \sigma v \rangle = \frac{\int F(v)\sigma(v)v dv}{\int F(v) dv},$$

where $F(v)$ is the Maxwell–Boltzmann speed distribution of the electrons.

High-density case: In this limit, $n_e \langle \sigma_{21} v \rangle \gg A_{21}$, and eq. (6.3) becomes

$$
\begin{aligned}
j_{21} &= n_e n_1 \frac{\langle \sigma_{12} v \rangle}{n_e \langle \sigma_{21} v \rangle} \frac{h\nu_{21}}{4\pi} \\
&= n_1 A_{21} \frac{h\nu_{21}}{4\pi} \frac{g_2}{g_1} e^{-\chi/kT_e}.
\end{aligned}
\tag{6.6}
$$

In the high-density case, collisional excitations are more likely to result in subsequent collisional de-excitations than radiative transitions, and eq. (6.6) shows that in this case $j \propto n$, since only a fixed fraction of the collisional excitations will lead to production of photons.

We can define a 'critical density' where the the radiative and collisional de-excitation rates are comparable, i.e.,

$$
n_{\text{crit}} = \frac{A_{21}}{\langle \sigma_{21} v \rangle} = \frac{A_{21}}{q_{21}}.
\tag{6.7}
$$

At densities around the critical value, the emissivity of a line goes from the $j \propto n^2$ regime to the $j \propto n$ regime. In the general case of a multi-level atom, the critical density is defined by equating all radiative de-excitations and collisional depopulations (i.e., excitation or de-excitation), i.e.,

$$
n_{\text{crit}} = \sum_{j<i} A_{ij} \bigg/ \sum_{j \neq i} q_{ij}.
\tag{6.8}
$$

Electric dipole transitions have spontaneous radiative transition rates of order $A \approx 10^8 \, \text{s}^{-1}$ and corresponding critical densities in excess of $10^{15} \, \text{cm}^{-3}$ for nebular temperatures. Non-electric dipole transitions, which have lower transition rates ('forbidden' lines that violate one of the quantum mechanical selection rules), are of more interest because the critical densities for these lines are in the range found in low-density nebulae (H II regions, planetary nebulae, and AGN narrow-line regions). Critical densities (primarily from De Robertis and Osterbrock 1986) for the strong forbidden lines in AGN spectra are given in Table 6.1.

Consider the pair of [S II] lines shown in Fig. 6.1. The critical density for the $^4S_{3/2}$–$^2D_{5/2}$ transition which leads to the $\lambda 6716$ line is $1.5 \times 10^3 \, \text{cm}^{-3}$, whereas the critical density for the $^4S_{3/2}$–$^2D_{3/2}$ transition which leads to the $\lambda 6731$ line is $3.9 \times 10^3 \, \text{cm}^{-3}$. Thus, for $n_e \ll 10^3 \, \text{cm}^{-3}$, their flux ratio is given by the ratio of their emissivities in the low-density regime (eq. 6.5),

$$
\frac{F(\lambda 6716)}{F(\lambda 6731)} \propto \frac{\langle \sigma_{\lambda 6716} v \rangle}{\langle \sigma_{\lambda 6731} v \rangle} \approx \frac{g_{\lambda 6716}}{g_{\lambda 6731}} = \frac{2(5/2)+1}{2(3/2)+1} = \frac{6}{4} = 1.5,
\tag{6.9}
$$

because the ratio of collisional excitation rates reduces only to the statistical weights of the levels since they lie so close in energy. Similarly, for $n_e \gg 10^3 \, \text{cm}^{-3}$, both lines are in the high-density limit, and their flux ratio, from eq. (6.6), is

$$
\frac{F(\lambda 6716)}{F(\lambda 6731)} \propto \frac{A_{\lambda 6716} \, g_{\lambda 6716}}{A_{\lambda 6731} \, g_{\lambda 6731}} \approx \left(\frac{2.6 \times 10^{-4} \, \text{s}^{-1}}{8.8 \times 10^{-4} \, \text{s}^{-1}} \right) \frac{6}{4} \approx 0.44.
\tag{6.10}
$$

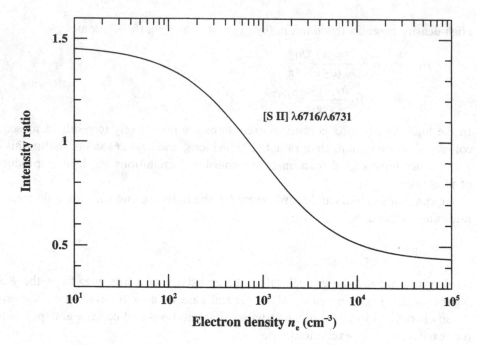

Fig. 6.2. The [S II] $\lambda6716$/[S II] $\lambda6731$ intensity ratio is a sensitive function of the electron density over the range $10^2 \, \text{cm}^{-3} \lesssim n_e \lesssim 10^4 \, \text{cm}^{-3}$ as depopulation of the upper levels changes from being primarily radiative at low densities to primarily collisional at high densities. The 'critical density' that marks the transition between these two regimes is a function of the Einstein coefficient A_{21} and cross-section for collisional de-excitation σ_{21}, which differ from line to line. The intensity ratio of these lines can be used to determine n_e as long as both lines are not in either the low or high density limit. Data courtesy of R. W. Pogge, based on atomic data from Cai and Pradhan (1993). The values given here are for $T_e = 10^4 \, \text{K}$.

In both cases, the ratio is set purely by atomic parameters.† However, at intermediate densities ($n_e \approx 10^3 \, \text{cm}^{-3}$) the $\lambda6716$ line is still in the low-density limit, but the $\lambda6713$ line is now becoming collisionally suppressed. Thus, the flux ratio of these two lines is *density dependent* over this range, since the ratio of their emissivities in this range is

$$\frac{j_{\lambda6716}}{j_{\lambda6731}} \propto \frac{n_e}{n_e^2} \propto n_e^{-1}. \tag{6.11}$$

Thus measurement of the flux ratio of these lines provides a measurement of n_e.‡ The theoretical relationship between the [S II] $\lambda6716$/[S II] $\lambda6731$ intensity ratio and electron density is shown in Fig. 6.2, which is consistent our approximate predictions (eqs. 6.9, 6.10, and 6.11) based on a simple three-level atom.

---

† We have ignored the very slight temperature dependence of the collisional excitation cross-sections in the low-density regime.

‡ The [S II] $\lambda6716$/[S II] $\lambda6731$ intensity ratio is *weakly* dependent on temperature as well, $\sim T_e^{1/2}$.

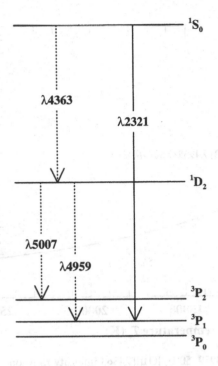

Fig. 6.3. A partial energy-level diagram for O^{++} showing the transitions that produce the O III $\lambda\lambda4364$, 4959, 5007 lines. The relative rate at which the $^1S_0$ and $^1D_2$ levels are populated by collisions is a sensitive function of temperature. Adapted from Osterbrock (1989), p. 64.

By using pairs of forbidden lines which arise out of different levels of the same multiplet, it is thus possible to measure densities from the flux ratio of the lines as long as one line is in the low-density limit ($j \propto n^2$) and the other is in the high-density limit ($j \propto n$). For the NLR in Seyfert galaxies, measured electron densities span the entire range measurable with the [S II] doublet, from $10^2\,\mathrm{cm}^{-3}$ to $10^4\,\mathrm{cm}^{-3}$. An 'average' NLR density is about $2000\,\mathrm{cm}^{-3}$ (Koski 1978).

6.2.2 Electron Temperatures

Again, to obviate complications due to unknown abundances and ionization fraction, it is desirable to measure the electron temperature from emission lines that arise in a single ion. This is done by using lines which have very different excitation potentials χ so that the rate at which the different levels are populated by collisions is highly temperature dependent. Suitable sets of lines for such measurements include [O III] $\lambda\lambda4363$, 4959, 5007 (see Fig. 6.3) and [N II] $\lambda\lambda5755$, 6548, 6583, although use of the latter in AGNs is hampered by the relative weakness of [N II] $\lambda5755$.

The flux ratio $F(\lambda4959 + \lambda5007)/F(\lambda4363)$ is very sensitive to the relative collisional

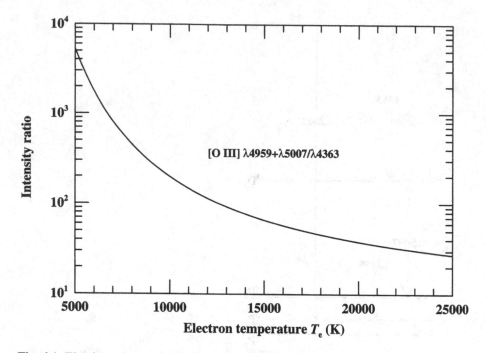

Fig. 6.4. The dependence of the [O III] $\lambda\lambda4959$, 5007/[O III] $\lambda4363$ intensity ratio on electron temperature T_e, as given in eq. (6.12) for negligible density ($n_e = 0$). Adapted from Osterbrock (1989), p. 120.

excitation rates of the $^1S_0$ and $^1D_2$ levels, and thus has a strong temperature dependence, which is given by

$$\frac{F(\lambda4959 + \lambda5007)}{F(\lambda4363)} \approx \frac{7.33 \exp\left(3.29 \times 10^4/T_e\right)}{1 + 4.5 \times 10^{-4}\, n_e T_e^{-1/2}} \qquad (6.12)$$

(Osterbrock 1989, p. 121). Equation (6.12) is shown graphically in Fig. 6.4.

The range of electron temperatures measured for the NLR in this way is about 10 000–25 000 K. A 'typical' value for the NLR is $T_e \approx 16\,000$ K (Koski 1978).

It is important to keep in mind that the NLR is characterized by a broad range of densities (see §6.4). The relative lines strengths depend on both n_e and T_e, so a measurement of T_e from the [O III] lines is not necessarily representative of the temperature of the [S II]-emitting clouds. Moreover, eq. (6.12) shows that the [O III] $\lambda4363$ emission may arise primarily in the denser clouds, which may not necessarily be representative of those that produce much of the [O III] $\lambda\lambda4959$, 5007 emission. Another important limitation of these methods of determining T_e and n_e is that they are determined from different ionization states. The values of the temperature and density are not expected to be the same, for example, in a region where the predominant ionization state of oxygen is O^+ as in a region where oxygen is primarily in the form O^{++}. More accurate determinations thus depend on *spatially resolved* spectrophotometry. The treatment described here is sufficient, however, for exploring the basic properties of the NLR.

6.3   Basic Parameters

As in the case of the BLR (Chapter 5), we can use the luminosity in strong lines to deduce some of the basic characteristics of the NLR. In the case of the NLR, the line of choice is the hydrogen Hβ recombination line, since to first approximation the narrow-line flux ratios are typical of Case B recombination, but with some slight modification of the hydrogen spectrum owing to collisional excitation out of the ground state (Gaskell and Ferland 1984). The advantage of using recombination lines is that unlike the collisionally excited lines, the strengths of recombination lines are relatively insensitive to T_e. However, the hydrogen Lyman series lines are optically thick (by definition in Case B), so the easily accessible Balmer lines must be used. We consider Hβ rather than Hα because it is somewhat less sensitive to collisional effects.

The emissivity of the gas in the Hβ line is simply (Osterbrock 1989, p. 78)

$$j_{H\beta} = n_e n_p \alpha_{H\beta}^{eff} \frac{h\nu_{H\beta}}{4\pi} = n_e^2 \alpha_{H\beta}^{eff} \frac{h\nu_{H\beta}}{4\pi} \tag{6.13}$$

$$= \frac{n_e^2}{4\pi} 1.24 \times 10^{-25} \text{ ergs s}^{-1} \text{ cm}^{-3} \text{ ster}^{-1},$$

where $\alpha_{H\beta}^{eff}$ is the effective recombination coefficient for Hβ† and we have assumed as an approximation a pure hydrogen, completely ionized gas, so the number of ions is $n_p \approx n_e$. The total luminosity in the Hβ line is given by integrating this emissivity over the entire emitting volume, and over all directions:

$$L(H\beta) = \int\int j_{H\beta} \, d\Omega \, dV = \frac{4\pi\epsilon n_e^2}{3} 1.24 \times 10^{-25} r^3 \text{ ergs s}^{-1}, \tag{6.14}$$

where as before ϵ is the filling factor, and as before the number of clouds N_c and cloud radius ℓ are related to the filling factor by $N_c \ell^3 = \epsilon r^3$.

By using eq. (6.14) and units appropriate for the NLR,

$$r \approx 19 \left(\frac{L_{41}(H\beta)}{\epsilon \, n_3^2} \right)^{1/3} \text{ pc}, \tag{6.15}$$

where $L_{41}(H\beta)$ is the NLR Hβ luminosity in units of 10^{41} ergs s^{-1}, and n_3 is the electron density in units of 10^3 cm^{-3}. Typical values of the Hβ luminosity in Seyfert 2 galaxies are 10^{39}–10^{42} ergs s^{-1}. In nearby AGNs, the NLR is often at least *partially* resolved, and the inferred sizes are generally $r \gtrsim 100$ pc, from which one can conclude that

---

† The effective recombination coefficient reflects the number of recombinations to *all levels* $n \geq 4$ which ultimately lead to an $n = 4 \rightarrow n = 2$ transition. The value of $\alpha_{H\beta}^{eff}$ depends only weakly on temperature, approximately as $T_e^{-0.9}$ (Osterbrock 1989, Table 4.2). At $T_e = 10^4$ K, $\alpha_{H\beta}^{eff} \approx 3 \times 10^{-14}$ cm^3 s^{-1}.

$$\epsilon \lesssim 10^{-2}, \tag{6.16}$$

and in some cases the filling factor must be several orders of magnitude below this limit. We can thus conclude that the NLR, like the BLR, is also clumpy. This conclusion is also borne out by the profile structure seen in high-resolution spectra. Moreover, the profile structure is seen to *change* as a function of position in partially resolved NLRs.

We can also estimate the mass of the NLR by using eq. (5.9),

$$M_{\text{NLR}} = \frac{4\pi}{3} \epsilon r^3 n_e m_p \tag{6.17}$$

and using eq. (6.14) and eq. (6.15) to write

$$\frac{4\pi \epsilon r^3}{3} = \frac{L(H\beta)}{n_e^2 \, 1.24 \times 10^{-25}} = 8.1 \times 10^{59} \frac{L_{41}(H\beta)}{n_3^2}. \tag{6.18}$$

Using this in eq. (6.17) gives

$$
\begin{aligned}
M_{\text{NLR}} &= 8.1 \times 10^{59} \frac{L_{41}(H\beta)}{n_3^2} n_e m_p \\
&= 1.4 \times 10^{39} \frac{L_{41}(H\beta)}{n_3} \text{ g} \\
&= 7 \times 10^5 \frac{L_{41}(H\beta)}{n_3} M_\odot.
\end{aligned}
\tag{6.19}
$$

It is thus seen by comparison with eq. (5.9) that the NLR is several orders of magnitude more massive than the BLR, although the amount of line emission it produces is often comparable to the BLR emission. The reason for this is, of course, that the emissivity of recombination lines is proportional to n_e^2, and the denser BLR is a much more efficient emitter.

It is also possible to infer some of the properties of the line-emitting clouds in the same fashion as in Chapter 5. For example, photoionization models of the NLR general give an ionization parameter (eq. 5.15) $U \approx 0.01$, so the Strömgren depth (eq. 5.18) is typically $r_1 \approx 10^{18} n_3^{-1}$ cm. If this is taken to be a lower limit on the size of the clouds ℓ, then it is easy to show that the number of narrow-line clouds is $N_c \lesssim 10^5 n_3^2 L_{41}(H\beta)$. We might therefore expect some clumpiness in the narrow-line profiles if indeed the number of clouds is small.

Another simple calculation that might be performed is to estimate the NLR covering factor from the equivalent width of Lyα, as we did with the BLR (§5.2). However, this calculation often fails spectacularly in the case of the NLR, yielding covering factors of only a few percent. The reason for this is again dust (Netzer and Laor 1993); the assumption that each ionizing photon leads to one Lyα photon is violated if Lyα photons are destroyed before they can diffuse out of the NLR clouds. Most Lyα photons are absorbed by dust before they can random-walk their way out of the clouds, and thus the Lyα equivalent width can thus grossly underestimate the NLR covering factor. On the other hand, there are sources (such as NGC 1068) where the continuum emission we observe is not a good measure of what is seen by the line-emitting clouds,

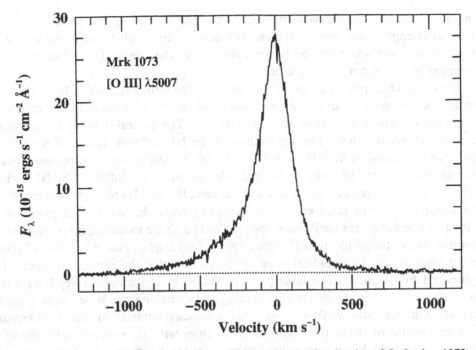

Fig. 6.5. The [O III] $\lambda 5007$ emission-line profile in the Seyfert 2 galaxy Markarian 1073 (Veilleux 1991) at a resolution of $10\,\mathrm{km\,s^{-1}}$. The line is asymmetric about the peak (which is defined to be $v = 0$), with more flux shortward than longward of line center. Data courtesy of S. Veilleux.

either because the continuum is emitted anisotropically or on account of attenuation along our line of sight to the nucleus. It is possible in such cases to deduce a covering factor greater than unity.

Finally, we note that narrow-line variability is in general not expected. Equation (6.15) implies that the light-travel time across the NLR is of order $\tau_{\mathrm{LT}} \approx 60\,(L_{41}(\mathrm{H}\beta)/\epsilon n_3^2)^{1/3}$ years, which means that only long-term continuum changes would show up as changes in the narrow-line fluxes. Furthermore, the long recombination time (eq. 5.33), $\tau_{\mathrm{rec}} \approx 130\,n_3^{-1}$ years, will damp out emission-line variations in the NLR. The conditions for narrow-line variability are thus (a) a higher than average electron density (say, $n_{\mathrm{e}} \approx 10^6\,\mathrm{cm^{-3}}$), (b) a compact NLR, and (c) a long secular trend in the continuum brightness. All of these conditions appear to be met in at least one case, the BLRG 3C 390.3 (Zheng *et al.* 1995b).

6.4 Narrow-Line Profiles

Narrow-line emission profiles are distinctly non-Gaussian; they have relatively stronger bases than a Gaussian, and they also tend to be asymmetric. A typical narrow-line profile is shown in Fig. 6.5. The profiles are usually blueward asymmetric, i.e., with more flux on the short-wavelength side of the line (relative to the peak) than on the long-wavelength side. Most of the excess blueward flux tends to be in the base of the line, so the cores are more nearly symmetric about their peak. This effect is relatively stronger in the higher ionization lines.

Since self-absorption cannot be important in the strong forbidden lines, the asymmetry clearly must be due to a combination of net radial motion of the clouds and some broad-band source of opacity, such as dust. The general sense of the asymmetry can be explained either (a) by net outflow of the NLR clouds through a dusty region (so that the emission from the clouds on the far, redshifted side is suppressed relative to the emission on the near, blueshifted side) or (b) by net infall of the NLR clouds which are themselves dust filled. In the latter case, the clouds radiate their line emission anisotropically, with relatively higher emissivity from the inward-facing side of the cloud. If one compares the Balmer-line profiles in a single spectrum, it is found that the profiles are virtually identical (Veilleux 1991). This implies that the source of opacity must have a very large optical depth ($\tau \gg 1$), and that the profile asymmetries are produced by obscuration of large parts of the NLR. If the optical depth were more modest ($\tau \approx 1$), then the effects of reddening and extinction would be comparable, and the $H\alpha/H\beta$ flux ratio would vary with radial velocity, contrary to what is observed.

Comparison of the narrow-line radial velocities with the systemic velocities of the host galaxies (which can be measured either from $H\,\textsc{i}$ 21-cm emission-line observations or from stellar absorption features in the host galaxy spectrum) shows that *centroids* of the narrow emission lines are often slightly blueshifted relative to the systemic velocities, typically by around 50–$100\,\mathrm{km\,s^{-1}}$ (Wilson and Heckman 1985). However, the observed range is fairly large, with the difference between the narrow-line and systemic radial velocities ranging from about $-250\,\mathrm{km\,s^{-1}}$ to $+250\,\mathrm{km\,s^{-1}}$. The *peaks* of the narrow emission lines, however, seem to be close to the systemic redshifts of the host galaxies. This indicates that most of the narrow-line emission arises in an approximately symmetric component at the systemic redshift and that there is a second weaker component that is slightly blueshifted relative to the systemic velocities, e.g., an outflowing component on the near side of the AGN.

The narrow-line profile widths are correlated with host-galaxy bulge luminosities. Through the Faber–Jackson relationship,† we assume that bulge luminosity is a direct

---

† For spheroidal systems, i.e., elliptical galaxies or spiral bulges, there is an empirical correlation, the Faber–Jackson relationship, between luminosity L and velocity dispersion σ that has the approximate form

$$L \propto \sigma^4$$

(Faber and Jackson 1976). The standard interpretation of this result is as follows: the luminosity is proportional to the central surface brightness Σ_0 and scale length r_s squared, i.e., $L \propto \Sigma_0 r_s^2$. Since Σ_0 shows little variation from galaxy to galaxy, $r \propto L^{1/2}$. Combining this with the virial theorem,

$$M\sigma^2 \propto \frac{M^2}{r},$$

and assuming that the mass-to-light ratio M/L is constant gives

$$\sigma^2 \propto \frac{M}{r} \propto \frac{L}{r} \propto \frac{L}{L^{1/2}} = L^{1/2},$$

or $L \propto \sigma^4$.

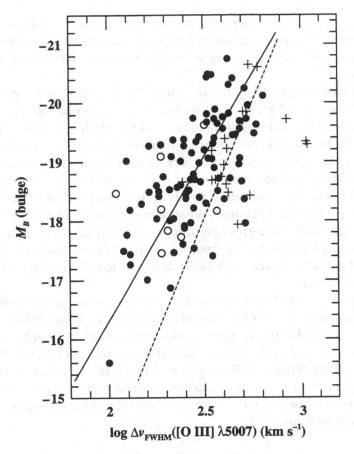

Fig. 6.6. The correlation between galactic bulge luminosity in the *B* band and the [O III] $\lambda5007$ emission-line width for Seyfert galaxies (filled circles). The open circles are low-luminosity ($L_\nu(1415\,\text{MHz}) < 10^{29.5}\,\text{ergs s}^{-1}\,\text{Hz}^{-1}$) radio sources with linear morphologies. The crosses are high-luminosity linear radio sources. The solid line is a least-squares fit to the filled circles, and the dashed line is the Faber–Jackson (1976) relationship for normal galaxies. This correlation suggests that the *stellar* bulge mass is the primary factor that determines the Doppler widths of the narrow emission lines. The tendency for the strong linear radio sources to lie to the right of either line suggests that acceleration of the narrow line-emitting gas by outflowing jets can also be important. Luminosities are calculated assuming $h_0 = 1$. Adapted from Whittle (1992).

measure of the bulge gravitational potential, which is also reflected in the line widths. The correlation between bulge luminosity and [O III] $\lambda5007$ line width is shown in Fig. 6.6. This indicates that the narrow-line widths are primarily virial in origin. However, unlike the BLR where the line widths appear to reflect the potential of the central object, the NLR widths reflect the gravitational field of the *stars*. It is notable that this relationship shows the least scatter when galaxies with linear radio structures (jets) are omitted. AGNs with jets lie off the general relationship, in the sense that the velocities are larger than expected. This may well indicate that there is a non-virial component to the velocities, arguing for the importance of shock interactions between the radio jets and the NLR gas.

The various forbidden-line profiles in the spectrum of a particular AGN are qualitatively similar, although they have different widths. This means that the emissivity of different lines varies in different ways with radial velocity, which presumably translates to different radial emissivity distributions for different lines. Measurement of the forbidden-line widths (Δv_{FWHM}) for a number of lines within a single spectrum shows a correlation between line width and both critical density n_{crit} and ionization potential (Pelat, Alloin, and Fosbury 1981, Filippenko and Halpern 1984, De Robertis and Osterbrock 1984, Espey *et al.* 1994). Although it is not clear whether n_{crit} or the ionization potential is more fundamental, the correlation with n_{crit} generally has higher statistical significance. Since Δv_{FWHM} seems to have a virial origin, these correlations imply radial stratification of the NLR of some sort, with either density or ionization level (or both) increasing towards the center.

If indeed both the density and velocity dispersion increase as we get closer to the nucleus, the NLR may merge more or less naturally with the BLR. Some weak support for this picture is provided by the apparent correlation between the widths of the narrow lines and the widths of the broad lines in Seyfert 1 galaxies in which both are measured (e.g., Cohen 1983, Whittle 1985b). Espey *et al.* (1994) show that in at least one object the forbidden lines ($n_{crit} \lesssim 10^8 \, cm^{-3}$) and semi-forbidden lines ($n_{crit} \gtrsim 10^9 \, cm^{-3}$) appear to follow a single line-width/critical-density relationship, which thus argues for continuity between the NLR and BLR.

6.5   Morphology of the Narrow-Line Region

As mentioned earlier, the NLR in nearby AGNs is often resolvable on the sky, revealing important information about the distribution and kinematics of the emission-line gas (e.g., Walker 1968). The number of sources in which the NLR morphology has been well studied has increased dramatically over the last few years as a direct result of the high-quality emission-line images made possible by corrected optics for the *HST* imaging cameras. Ground-based emission-line imaging has also undergone something of a renaissance as a result of the availability of large-format, high-quality CCDs and improvements in narrow-band imaging technology, such as the emergence of commerically available high-stability Fabry–Perot etalons for high spectral-resolution imaging spectroscopy.

An emission-line image of an extended source is produced by obtaining at least two narrow-band images of the source, with the bandpass of one filter centered on an emission line (the 'on-band image') and the bandpass of the second filter centered on an adjacent 'line-free' region of the continuum (the 'off-band image'). A pure emission-line image is obtained by subtracting the off-band image from the on-band image, which removes all the continuum radiation that is in both images and leaves only the emission-line surface-brightness distribution. In practice, this is a process that requires great care, but can produce emission-line images of high photometric accuracy (Pogge 1992). With multiple photometric images, we can divide one emission-line image by another to produce maps of emission-line ratios, which can be used as spatially resolved diagnostics not only of density and temperature (§6.2), but excitation and ionization as well, by comparison with the BPT diagrams discussed in Chapter 2. For example, the [O III] $\lambda5007/H\alpha$ ratio can be used as an ionization diagnostic.

In general, the NLR morphology is found to be axisymmetric rather than spherically symmetric. The NLR axis coincides with the radio axis in sources where extended radio emission is detected, which indicates that there is some connection between the thermal narrow-line gas and the non-thermal plasma, as already suggested by larger than usual emission-line widths in AGNs with linear radio structures. In some sources, the radio morphology shows rather clear evidence for shock fronts at the interface between the radio-emitting plasma and the interstellar medium in the host galaxy. Strong narrow-line emission is observed from the post-shock cooling regions. The picture that emerges is that the outflowing plasma that is responsible for the radio emission creates bowshocks as it collides with the ambient NLR gas. While the post-shock gas may be primarily photoionized by the central source, collisional ionization also plays a rôle, and indeed composite ionization models seem to be necessary to match the NLR spectra accurately (e.g., Viegas-Aldrovandi and Contini 1989). The large narrow-line widths in Seyfert galaxies with linear radio sources are attributable to acceleration of the ambient gas by the radio-emitting plasma (e.g., Pedlar, Dyson, and Unger 1985); if there is an obscuring medium in the midplane (see below and Chapter 7), then the near-side, outflowing post-shock gas contributes a relatively blueshifted component to the narrow-line profiles, and at least qualitatively accounts for the narrow-line asymmetries (§6.4). Finally, ionizing radiation can be produced by shocks, thus 'auto-ionizing' the post-shock gas (Sutherland, Bicknell, and Dopita 1993) and providing an extra source of energy.

The most dramatic morphological features seen in AGNs are 'ionization cones' (e.g., Pogge 1988, Tadhunter and Tsvetanov 1989) which stand out clearly in maps of high-excitation lines such as [O III] $\lambda5007$ (Fig. 6.7). Ionization cones have been detected in somewhat more than a dozen AGNs, primarily Seyfert 2 galaxies, and thus far only in radio-quiet objects (Wilson and Tsvetanov 1994). These are wedge-shaped structures with opening angles typically in the range $\sim 30°-100°$. Inside the cone, the [O III] $\lambda5007/H\alpha$ flux ratio is greater than unity, which is characteristic of low-density gas ionized by an AGN continuum, whereas outside the cone this ratio is lower, indicating that this gas is ionized primarily by starlight rather than by the AGN. The sizes of ionization cones are in the range $\sim 50\,h_0^{-1}$ pc to $\sim 15\,h_0^{-1}$ kpc, and constitute what is sometimes referred to as the 'extended narrow-line region' (ENLR), in contrast to the higher surface-brightness, high-ionization gas closer to the nucleus, the 'classical NLR'. The edges of the cones are sharp and linear in excitation maps, which implies that the regions are defined by collimation of light from the nuclear source, not by

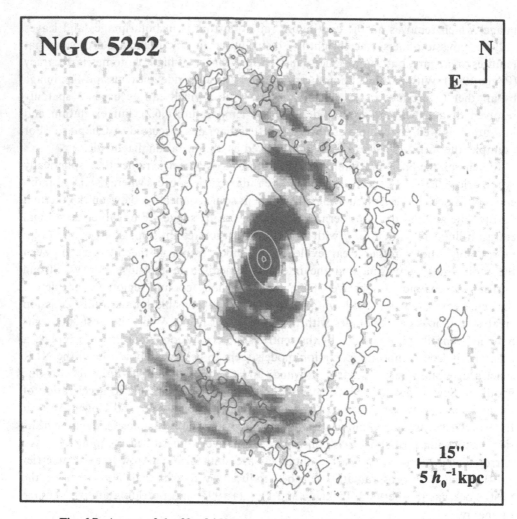

Fig. 6.7. A map of the [O III] λ5007 line emission in Seyfert 2 galaxy NGC 5252 (Tadhunter and Tsvetanov 1989, Wilson and Tsvetanov 1994) shows that the ionized region is confined to broad wedge-shaped regions ('ionization cones') on either side of the nucleus. The contours show the surface-brightness distribution of starlight. Data courtesy of Z. Tsvetanov, figure composed by R. W. Pogge.

the distribution of gas. Further evidence for this is provided by an H I 21-cm map of NGC 5252, which shows clearly that the *neutral* gas apparently avoids the region defined by the ionization cone (Prieto and Freudling 1993). The gas velocities in the ENLR seem to be consistent with rotation within the galaxy, without acceleration by interaction with radio-emitting plasma as in the classical NLR. The ENLR emission can be plausibly attributed simply to interstellar gas which is photoionized by nuclear source.

Ionization cones can be either single-sided or bi-conical structures, with the bi-conical structures always sharing a common axis. The axis does not seem to be correlated with the host-galaxy symmetry axis, however. Presumably in the case where a single-sided structure is detected, we do not detect the second cone because it is obscured from our view by dust in the plane of symmetry, which is required in any case to explain the narrow-line profiles.

Ionization cones lead us to an important realization, namely that the ionizing continuum is being emitted anisotropically, or at least that the radiation is anisotropic on kiloparsec scales. It is not clear whether the anisotropy is intrinsic to the continuum source, or if the anisotropy is introduced on scales smaller than the NLR, perhaps by the obscuring torus that is invoked in AGN unification models (Chapter 7); the apex of ionization cones is often dark, which suggests that the cone emerges through obscuring material that blocks our direct view of the nucleus in these sources. It is not clear, however, whether the torus itself collimates the ionizing radiation, or alternatively if the nuclear radiation is already collimated at smaller radii and the hole in the torus is caused by the beamed radiation.

7   Unified Models of AGNs

In the last four chapters, we have examined the individual components that constitute AGNs. At this point, it is worth briefly summarizing these as we proceed to develop a more global picture of the AGN phenomenon. Although direct proof is lacking, the evidence points towards gravitational accretion of matter by supermassive black holes as being the primary energy source in AGNs. Gravitational potential energy is converted into radiation via viscous dissipation in an accretion disk surrounding the black hole. For a luminous Seyfert galaxy, the black-hole mass is inferred to be something like $\sim 10^7 M_\odot$ (i.e., $R_S \lesssim 10^{13}$ cm), and the UV/optical continuum-emitting region of the purported accretion disk is smaller than $\sim 10^{15}$ cm. The corresponding X-ray-emitting region appears to be smaller still, perhaps only several times R_S. Surrounding this is the BLR, which has a typical size of 10^{16} cm or so, but whose specific geometry and kinematics are poorly known. It appears that most of the IR continuum emission arises on spatial scales larger than the dust sublimation radius ($\gtrsim 10^{17}$ cm). This is also where we find the lower-density NLR, where the gas motions are dominated by gravity, but where we also see evidence for interaction with jets in the form of shock-heating and outflowing gas. Even in the nearest AGNs, the NLR is the smallest scale on which details can be spatially resolved in the UV through IR spectrum, and on such scales ($\gtrsim 10^{18}$ cm) there is clear evidence for anisotropy – AGNs appear to have axial rather than spherical symmetry, and it is hypothesized that all of the unresolved AGN components are surrounded by an optically thick obscuring torus that permits the AGN radiation to escape only along the torus axis, which is defined by large-scale ionization cones. In those AGNs that are radio sources, the radio axis apparently aligns with the torus axis. The torus/radio-source axes do not seem to show any preferential orientation relative to the rotation axis of the host galaxy.

The clear evidence for anisotropy of both radio emission and higher-frequency radiation implies that the appearance of a given AGN will depend strongly on the observer's location relative to the axis of symmetry. Indeed the observed characteristics of a particular AGN might be so strongly orientation dependent that the classification of the system might be a function of the viewing angle. This is the fundamental notion behind what are generally referred to as 'unified models' of AGNs; the basic motivation for consideration of such models is that they appeal to our belief that any description of nature must be made as simple as possible in the absence of evidence to the contrary (the principle of Occam's razor). In this chapter, we will begin to address the question of the relationships among different types of AGNs and consider in at least a cursory fashion how much of the variety seen in the AGN family can be attributed to orientation effects.

7.1 Unification Ideas and Principles

Much of the work on unified models is morphological in nature, i.e., based on searches for correlations among observed parameters, which are motivated by some broad ideas about what kinds of physical effects might be operating. The goal of morphological studies is to try to find the minimum number of parameters needed to specify the problem and to determine the relationships among various parameters in the hope that this will lead to physical insight. The assumption is that there is less *intrinsic* diversity among AGNs than we observe, and that the wide variety of AGN phenomena we see is due to a combination of real differences in a small number of physical parameters (like luminosity) coupled with *apparent* differences which are due to observer-dependent parameters (like orientation).

Unified models of AGNs can be characterized as either 'strong' or 'weak', depending on the number of fundamental parameters allowed. Weak unification models allow more physical diversity and attempt to explain the relationships among a limited number of AGN types. An example of a weak unification model is one that allows two intrinsic parameters, optical and radio luminosity. In these models there are two basic types of AGNs, radio-quiet and radio-loud. In each type, we see a wide range of phenomena that have to do with variations in these two basic parameters, plus apparent differences due to the orientation of the system relative to the observer. The complementary strong model, on the other hand, assumes that there is only a single intrinsic parameter, the total luminosity (i.e., the optical and radio luminosity are correlated), and that all of the differences we observe, including the differences in the optical and radio properties, are ascribable to various orientation effects.

Historically there have been many attempts to develop unified models in order to explain the wide variety of AGN types (Chapter 2) in terms of a limited number of fundamental physical phenomena. Optical unification schemes have their origin in the question of the nature of Seyfert 2 galaxies – why are there AGNs without broad emission lines? Recognition that this might be an orientation effect involving obscuration of the central regions goes back at least as far as Osterbrock (1978). In the case of radio sources, conventional synchrotron models predict a high degree of anisotropy. Blandford and Rees (1978) were the first to recognize that blazar phenomena (rapid variability, high brightness temperatures, high polarization) could be accounted for by supposing that these were otherwise 'normal' AGNs which we are viewing along the radio axis and thus the observed flux is dominated by the beamed component.

For illustrative purposes, we summarize some basic aspects of a general unification scheme in Table 7.1. The key elements in this simple scheme are the obscuring torus and, in the case of the radio sources, a synchrotron-emitting jet. In this (weak)

Table 7.1
Possible Simple Unifications

Radio	Orientation	
Properties	Face-On	Edge-On
Radio Quiet	Seyfert 1	Seyfert 2
	QSO	FIR galaxy?
Radio Loud	BL Lac	FR I
	BLRG	NLRG
	Quasar/OVV	FR II

scheme, Seyfert 1s and BLRGs are distinguished from their counterparts, Seyfert 2s and NLRGs, by the orientation of the obscuring torus. If the torus is seen face-on, our view of the central regions is unobstructed and we detect the broad lines; if our view is closer to edge-on, the central regions are not seen directly and no broad lines are detected. BL Lacs and OVVs are both face-on versions of radio sources (i.e., the radio-source axis parallel to the line of sight, or 'pole-on'), with the BL Lacs corresponding to the low-luminosity sources (FR I) and the OVVs corresponding to the more luminous sources (FR II).

Figure 7.1 illustrates the basic concept of simple unification models, and shows in a qualitative way how AGN classification might be aspect dependent.

We emphasize that at this stage it is not clear what the fundmental parameters in a unification scheme should be. While luminosity and orientation provide us with a good starting point, there may be other important factors such as the galaxy morphology and gas/dust content. It is doubtful that the geometry of the obscuring torus is the same in all AGNs, and indeed it may be related to the AGN luminosity. Finally, when we are comparing objects at different redshifts, evolution of the sources on cosmological time scales (Chapter 11) could be important.

7.2   Evidence for Unification in Seyferts

The first basic question we need to address is why there are two types of Seyfert galaxies. From an observational point of view, the differences are (a) that Seyfert 2 galaxies lack the broad emission lines seen in Seyfert 1 galaxies, and (b) that the AGN (or 'featureless') continuum is weaker relative to the stellar continuum in Seyfert 2s than in Seyfert 1s. An early suspicion was that in fact there is intrinsically only one type of galaxy (Seyfert 1), but in some cases we observe the nuclear regions through an attenuating medium that partially extinguishes the continuum and the broad lines. Since the attenuation must be over a broad wavelength range, dust is the obvious

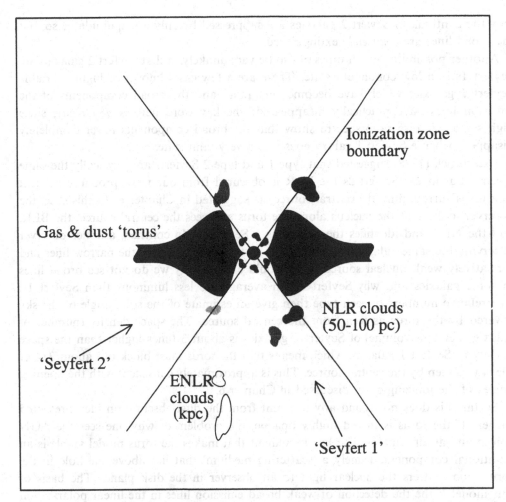

Fig. 7.1. A conceptual scheme for unification of Seyfert 1 and Seyfert 2 galaxies, not to scale. A highly opaque dusty torus surrounds the continuum source (black dot at center) and broad-line region (clouds near center). These cannot be viewed directly by an observer close to the torus midplane, although the narrow-line region and extended narrow-line region can be seen directly. This would lead the observer to classify the galaxy as a 'Seyfert 2'. An observer closer to the torus axis would have an unobscured view of the nuclear regions and classify the same galaxy as a 'Seyfert 1'. Figure courtesy of R. W. Pogge.

candidate. The problems with this explanation are first, that the featureless continuum in Seyfert 2s still looks more or less like a power law, which it should not if in fact it is heavily reddened; a reddened power law is no longer a power law. The UV spectra of Seyfert 2 galaxies have approximately the same shape as the UV spectra of Seyfert 1 galaxies (Kinney *et al.* 1991). And second, the typical Seyfert 2 galaxy is only about one magnitude fainter than the typical Seyfert 1 galaxy. Thus, it is hard to understand why the continua in Seyfert 2 galaxies are suppressed by only a magnitude or so, but the broad lines are *completely* extinguished.

Another possibility, which turns out to be very unlikely, is that Seyfert 2 galaxies are Seyfert 1s in a low continuum state. There are a few cases known of highly variable Seyfert 1 galaxies which have become very faint, and the broad components of the emission lines have practically disappeared; the key word here is *practically*, since high signal-to-noise ratio spectra show that the broad components never completely disappear when a Seyfert 1 galaxy goes into a very faint state.

Osterbrock (1978) suggested that type 1 and type 2 Seyferts are physically the same objects, but in the Seyfert 2s the BLR is obscured from our view, probably by dust in a torus surrounding the central source, as suggested in Chapter 6. In this case, the observer looking at the nucleus along the torus axis sees the central source, the BLR, and the NLR and identifies the galaxy as a Seyfert 1. In contrast, another observer observing the same galaxy but in the plane of the torus sees only the narrow lines and a relatively weak nuclear source. This would explain why we do not see broad lines in some galaxies and why Seyfert 2s on average are less luminous than Seyfert 1s. The relative numbers of each type then give an estimate of the solid angle of the sky covered by the torus as seen from the central source. The space density (number of galaxies per unit volume) of Seyfert 2 galaxies is about 3 times higher than the space density of Seyfert 1 galaxies, which means that the torus must block out about 3/4 of the sky as seen by the central source. This is approximately consistent with the opening angles of the ionizing cones discussed in Chapter 6.

So far this does not sound any different from the dust-obscuration idea presented earlier. If the torus is indeed highly opaque, the problem of why one sees the AGN continuum at all remains. The key ingredient that makes the torus model viable is an additional component, namely a 'scattering medium' that lies above the hole in the torus, and scatters the nuclear light to an observer in the disk plane. The basis of this model is the the detection of weak broad emission lines in the linear polarization spectrum of the Seyfert 2 galaxy NGC 1068 (Fig. 7.2) and other sources (e.g., Antonucci and Miller 1985). In NGC 1068, the continuum polarization is $\sim 16\%$, after removal of the significant starlight contribution. A polarized spectrum can result from scattering or reflection of the AGN continuum, either by dust or by free electrons. The polarization of the featureless continuum in NGC 1068 is wavelength independent as far into the UV as $\sim 1500\,\text{Å}$ (Code *et al.* 1993), which indicates that the scattering particles are electrons rather than dust. The polarization **E** vector is perpendicular to the radio axis of the source, as would be expected in a single-scattering model† (Fig. 7.3). The narrow lines, in contrast to the other AGN spectral features, are polarized by $\lesssim 1\%$ since the NLR is observed directly and not after scattering. Thus, the nuclear components of the spectrum are weak and polarized, but are unreddened.

---

† The polarization position angle is perpendicular to the photon's direction of flight prior to scattering. If the photons were scattered several times prior to heading towards the Earth, their flight paths immediately prior to the final scattering would be random, and no net polarization would be observed.

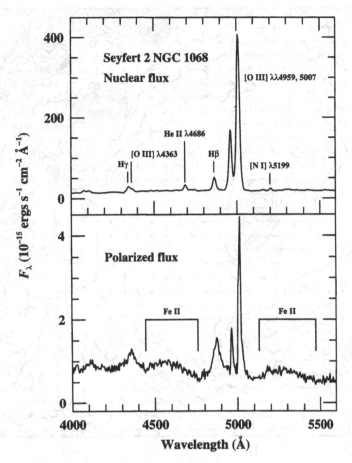

Fig. 7.2. The top panel shows the total-flux nuclear spectrum of the Seyfert 2 galaxy NGC 1068, and the bottom panel shows the linearly polarized flux (Miller, Goodrich, and Mathews 1991). The polarization spectrum is that of a Seyfert 1 galaxy; the Balmer lines are broad, and Fe II blends, which are characteristic of broad-line but not narrow-line spectra, are seen. NGC 1068 is apparently has a Seyfert 1 nucleus that we do not view directly, but only in scattered (and thus polarized) light. Data courtesy of R. W. Goodrich.

The NGC 1068 observations demonstrate that there are at least *some* Seyfert 2 galaxies that are intrinsically Seyfert 1s. Other similar cases have also been found among Seyfert 2s with polarized continua (e.g., Tran, Miller, and Kay 1992). There remain many difficulties in interpretation of these observations, including the fact that Seyfert 2 galaxies sometimes show an *extended* component of continuum emission whose origin is unknown (e.g., Pogge and De Robertis 1993, Antonucci, Hurt, and Miller 1994, Tran 1995), but which may be related to extended soft X-ray emission. Nevertheless, the evidence that some Seyfert 2 galaxies have hidden Seyfert 1 properties

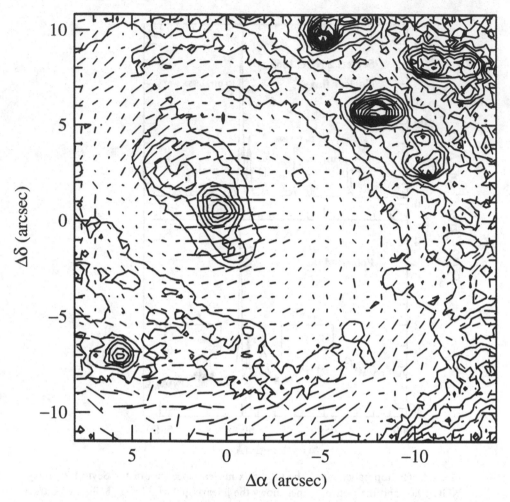

Fig. 7.3. The contours show a color map of the nuclear regions of the Seyfert 2 galaxy NGC 1068, constructed by taking the ratio of near-ultraviolet ($\sim 3600\,\text{Å}$) to red ($\sim 6100\,\text{Å}$) flux; the highest-level contours correspond to the bluest regions. Superposed on the color map is a map of the polarization at 4260 Å; the amplitude of the linear polarization vector is indicated by the length of each line segment, and the orientation of each line element indicates the position angle. The coordinate system is centered on the NGC 1068 nucleus, with north up and east to the left. The bluest regions (mostly in the upper right), which are compact and unpolarized, are stellar OB associations. The extended very blue regions are the most highly polarized; this is due to scattered nuclear light. Note how the orientation of the polarization vectors in the highly polarized regions around ($\Delta\alpha \approx +3''$, $\Delta\delta \approx +3''$) and near the bottom of the figure are oriented perpendicular to the line to the origin, as expected in a single-scattering model. Figure from Pogge and De Robertis (1993), courtesy of R. W. Pogge. Polarization data are from Miller, Goodrich, and Mathews (1991). A similar map based on *HST* data is shown by Capetti *et al.* (1995).

is so strong that we need to ask whether or not it is possible that *all* Seyfert 2s are really Seyfert 1s with obscured BLRs.

One possible way to assess the anisotropy of the radiation field in Seyfert 2s is to sum up the energy in the emission lines arising in the ionization cone, and compare the total energy with that expected from the nucleus as we observe it (e.g., Wilson, Ward, and Haniff 1988). The way that this is usually done is to assume that the gas in the ionization cone is more or less smoothly distributed and ionized by the central source. The luminosity in the Hβ line, from eqs. (6.13) and (6.14), is

$$L(\text{H}\beta) = \int\int j_{\text{H}\beta} \, d\Omega \, dV = \alpha_{\text{H}\beta}^{\text{eff}} \, h\nu_{\text{H}\beta} \int n_{\text{e}}^2 dV. \tag{7.1}$$

Assuming ionization equilibrium, the number of ionizing photons produced by the central source (eq. 5.14) is equal to the total number of recombinations over the ionized volume (cf. eq. 5.17), so

$$Q(\text{H}) = \int_{\nu_1}^{\infty} \frac{L_\nu \, d\nu}{h\nu} = \alpha_{\text{B}} \int n_{\text{e}}^2 \, dV. \tag{7.2}$$

We can combine these to infer the number of ionizing photons produced per second from the Hβ luminosity,

$$Q(\text{H}) = \frac{L(\text{H}\beta)}{h\nu_{\text{H}\beta}} \frac{\alpha_{\text{B}}}{\alpha_{\text{H}\beta}^{\text{eff}}} \approx 2.1 \times 10^{53} \, L_{41}(\text{H}\beta) \text{ photons s}^{-1}. \tag{7.3}$$

This can be compared with the ionizing photon production rate that we would infer directly from the observed continuum (cf. eq. 5.10) and assuming isotropic emission, i.e.,

$$Q_{\text{obs}}(\text{H}) = 4\pi d^2 \int_{\nu_1}^{\infty} \frac{F_\nu \, d\nu}{h\nu} = 4\pi d^2 \, \Phi(\text{H}), \tag{7.4}$$

where d is the distance to the AGN. This calculation usually reveals that $Q(\text{H})/Q_{\text{obs}}(\text{H}) \gtrsim 1$, which suggests that the gas in the ionization cone sees a more luminous continuum than we do, thus supporting the basic unification picture. Unfortunately, this calculation is not particularly robust. If instead of a continuous distribution of gas we consider a discrete-cloud model (i.e., a system of optically thick clouds in a region with a filling factor $\epsilon < 1$ as we did in §6.3, we replace eq. (7.1) with eq. (6.14) and then eqs. (5.15) and (5.18) give

$$Q(\text{H}) = 4\pi r^2 \, n_{\text{e}}^2 \, \alpha_{\text{B}} \, r_1, \tag{7.5}$$

where r_1 is the Strömgren depth of an individual cloud (eq. 5.18). We thus obtain

$$Q(\text{H}) \approx \frac{L(\text{H}\beta)}{h\nu_{\text{H}\beta}} \frac{\alpha_{\text{B}}}{\alpha_{\text{H}\beta}^{\text{eff}}} \frac{r_1}{r} \epsilon^{-1}. \tag{7.6}$$

The ratio $Q(\text{H})/Q_{\text{obs}}(\text{H})$ thus depends on the filling factor, the size of the emitting region, and the properties of the individual clouds.

A similar test is to compare the infrared luminosity to the luminosity at higher energies (X-ray through $\sim 1\mu$m). The assumption is that the torus absorbs the high-energy radiation and reprocesses it into IR radiation. The total IR luminosity of the torus is thus equal to the total higher-energy luminosity it intercepts. This calculation is subject to many significant uncertainties even when it is done carefully (e.g., Storchi-Bergmann, Wilson, and Baldwin 1992), with the most important problem being our inability to isolate the nuclear from the extended far-IR flux with the poor spatial resolution of the *IRAS* data.

While these tests are not conclusive, they do support the basic unification idea, namely, that some AGNs that we call Seyfert 2s are actually more luminous Seyfert 1s observed at a disadvantageous angle.

7.3   Statistical Tests for Unification

Before continuing with a discussion of the possible unification of various types of objects, it is worth pausing to consider the general nature of observational tests that can be carried out to assess the viability of possible unification models. The tests described previously apply to individual objects – what we would like to know is whether or not we might be able to account for the general population of certain types of objects (like Seyfert 2s) on the basis of orientation effects.

We assume that we are starting out with a group of similar objects, the 'parent population', whose observed properties differ depending on how they are oriented relative to us. We assume, of course, that averaged over a sufficiently large number of such objects their orientation as seen by us is random. A first very obvious criterion for unification is that the space density of the observed objects must be smaller than the space density of the parent population. Very rough current space densities for some of the types of object in Table 7.1 are given in Table 7.2 (from Woltjer 1990). Also, as we will discuss in Chapter 11, some parent populations are known to change with time.† Again, obviously, the selected objects and the assumed parent population must evolve in a consistent manner.

Finally, it is important to keep in mind that in attempting to ascribe the differences between samples of AGNs to orientation effects, one must have some evidence that they are in fact intrinsically similar (i.e., drawn from the same parent population). For example, we cannot blindly compare the properties of all known Seyfert 1 galaxies to all known Seyfert 2 galaxies because we do not know *a priori* that the distribution in intrinsic luminosity is the same. Indeed, it is one of the tenets of unification that galaxies with the same luminosity will appear to have very different *inferred* luminosities because their continua are not radiated isotropically. Thus, in comparing samples of different types of objects, one must pay close attention to how the objects are selected.

---

† When we see objects at large redshift, we see them as they were in the distant past because of the large light-travel times involved. By measuring the space density of a particular object nearby and comparing to the space density of similar objects at large redshifts, we can look for evolution of the population on cosmological time scales (Chapter 11).

Table 7.2

**Current Space Densities
of Selected Objects**

Type of Object	Space Density (Gpc^{-3})
Radio Quiet:	
Sy 2	$8 \times 10^5 h_0^3$
Sy 1	$3 \times 10^5 h_0^3$
QSOs	$800 \, h_0^3$
Radio Loud:	
FR I	$2 \times 10^4 h_0^3$
BL Lacs	$600 \, h_0^3$
FR II	$80 \, h_0^3$
Quasars	$20 \, h_0^3$

For Seyfert galaxies, one obvious way to formulate a comparison is to use a *distance-limited* sample of Seyfert 1 and Seyfert 2 galaxies; i.e., by considering all AGNs within some fixed volume, one should be able to compare with confidence the observed properties of all different types of galaxies within the volume.

Another way to proceed is to compare only those objects for which there is some independent indicator of the luminosity that does not depend on viewing angle. In other words, we seek an AGN component that emits radiation isotropically and is free of aspect-dependent obscuration, and that can be assumed to provide some indication of the total level of AGN activity in the galaxy. We consider several possibilities:

Extended radio emission: In Chapter 1, we noted that, in general, radio sources consist of two components, an extended steep-spectrum component and a compact flat-spectrum core. It is usually assumed (but with incomplete justification) that the extended component, which usually has a double-lobed structure centered on the compact component, is radiated isotropically. The core emission arises in a jet-like structure and is clearly highly beamed. For a given intrinsic ratio of core-to-lobe flux and distribution of Lorentz factors for the compact source, the *observed* core-to-lobe flux ratio R depends only on the angle of the jet relative to the line of sight. It is thus assumed that statistically R is a measure of orientation (e.g., Orr and Browne 1982). A homogeneous sample of radio sources thus should be characterized by comparable extended rather than total radio luminosity. In considering possible unification schemes, we must distinguish carefully between 'core-dominated' radio sources (high R sources in which the beamed component dominates), and 'lobe-dominated' sources (low R sources where the beamed component is not close to the observer's line of sight). Since the spectra of the two components are in general markedly different, it is statistically equivalent to distinguish between 'flat spectrum' and 'steep spectrum' radio sources.

Hard X-ray emission: Since the primary source of opacity in the hard X-ray spectrum is electron scattering, very large column densities, $N_e > 1/\sigma_e \approx 10^{24}\, \mathrm{cm}^{-2}$, are required to attenuate X-ray emission. Lawrence and Elvis (1982) showed that Seyfert 2 galaxies have significantly lower hard X-ray luminosities than Seyfert 1 galaxies, and in fact proposed a Seyfert unification scheme based on this, supposing that the Seyfert 2s are in fact highly attenuated Seyfert 1s.

Far-infrared emission: It is sometimes assumed that the far-IR radiation from galaxies is emitted isotropically. However, the obscuring torus is inferred to have a very high opacity (Krolik and Begelman 1988) and even in the far-IR the torus spectrum is expected to be strongly aspect dependent (Pier and Krolik 1992), and this seems to be supported by observations (Heckman, Chambers, and Postman 1992). Furthermore, the X-ray flux from the central source results in the torus being partially ionized (with an ionized fraction of $\sim 10^{-3}$), which means it also has significant free–free opacity at low radio frequencies (Krolik and Lepp 1989).

Extended narrow-line emission: In the optical, for example, one might consider the total luminosity in the *extended* emission-line regions, which should in principle not be affected by the obscuring torus which surrounds the central source. However, dust within the host galaxy presents a serious problem.

Thus, assembling samples of possibly related types of AGNs that span similar ranges in intrinsic luminosity is extraordinarily difficult since isotropic emission cannot be assumed for any waveband. Only the *extended* radio component in radio-loud objects can be reasonably assumed to be emitted isotropically, and then we must keep in mind that the extended radio emission reflects the state of the active nucleus as much as millions of years in the past, not at the current epoch.

7.4   Results of Unification Studies

7.4.1   Radio-Quiet Objects

There are several known Seyfert 2 galaxies whose polarization spectra reveal the presence of hidden Seyfert 1s. In the initial searches for such objects, the detection rate was approximately 50%, although this fraction might not be representative because the candidate sources were selected on the basis of high broad-band polarization. Whether or not *all* Seyfert 2s are intrinsically Seyfert 1s is an unsettled issue. Arguments in favor of Seyfert unification include:

(1) At least *some* Seyfert 2s are undoubtedly Seyfert 1s in which we observe the Seyfert 1 spectrum only in the scattered component of the spectrum.

(2) The narrow-line properties of Seyfert 1s and Seyfert 2s are statistically indistinguishable (Cohen 1983, Whittle 1985a), although Cohen (1983) notes that very high-ionization narrow lines tend to be found only in Seyfert 1 spectra. There is no reason to believe that these narrow-line spectra are drawn from different parent distributions. The situation may be different at higher luminosities; Wills *et al.* (1992) find that the narrow components of the UV lines are anomalously weak in radio-loud quasars compared to Seyfert 2s.

On the other hand, there are several lines of evidence that suggest that not all Seyfert 2s harbor hidden Seyfert 1s, specifically:

(1) The continua of Seyfert 2 galaxies are in general *not* polarized. This could suggest only the absence of the scattering medium in these sources, not the absence of a BLR, or that multiple (depolarizing) scatterings occur. A blue featureless continuum is still detected, despite the absence of polarization and broad lines. Multiple scatterings introduce other complications, such as broadening of the scattered lines.

(2) There are no known very high-polarization Seyfert 2s. In an edge-on system, continuum polarizations could be expected to reach 50% (Miller and Goodrich 1990), far higher than any known source.

(3) All QSOs have Type 1 spectra. If unification is correct, then where are the Type 2 QSOs? Possible speculative answers are either (a) that the obscuring tori do not occur in higher-luminosity sources, perhaps because the more intense radiation field destroys the dust, or (b) that perhaps the tori in high-luminosity sources are thinner. In either case, the BLR is less obscured in high-luminosity sources. Another possibility is that such objects exist, but are not immediately recognizable as AGNs. There are indeed several extraordinarily luminous *IRAS* sources, sometimes called 'far-infrared (FIR) galaxies' that have quasar-like luminosities and narrow-line spectra (e.g., Jannuzi *et al.* 1994).

7.4.2 Radio-Loud Objects

Radio-loud sources consist of two components, a presumably isotropically emitting, steep-spectrum extended source, which generally has a double-lobed structure, and a highly beamed flat-spectrum core, which is often seen to have a linear or jet structure at high angular resolution. Unlike the extended components, the jet structures are almost always one-sided, but co-aligned with the axis of the extended radio lobes. The usual explanation for the one-sidedness of the jets is that their radiation is relativistically beamed with fairly large Lorentz factors, which strongly favors detection of the approaching jet over the receding (or counter) jet. Evidence that supports this picture is that the extended emission on the jet side tends to show lower Faraday depolarization than the extended emission from the opposite side (Garrington *et al.*

1988, Laing 1988). As electromagnetic radiation of wavelength λ propagates through an ionized medium of electron density n_e in which there is a magnetic field, the plane of polarization is rotated by an angle

$$\Delta\phi \propto \lambda^2 \int n_e(x)B_\parallel(x)\,dx, \tag{7.7}$$

where $B_\parallel$ is the magnetic field component parallel to the Poynting vector and the integral is over the path through the medium. The galaxies appear to be surrounded by a diffuse ionized medium with randomly oriented magnetic fields, and different amounts of Faraday rotation along slightly different sight lines, and thus different electron column densities, through such a medium tend to depolarize radio emission. This effect is greater for a longer path length through the medium, so we can infer that the more depolarized radio lobe is on the far side of the galaxy.

Radio-loud quasars have presented a number of difficulties that might be overcome by unification models. The first of these is that if the simple two-component description given above is approximately correct and the jet axes are randomly oriented relative to the observer, we should often see sources in which the jet axis is close to the plane of the sky, in which case both the jet and counter-jet should have about the same surface brightness and should be equally detectable, in contradiction to the observations. Second, another aspect of the same problem is that if quasars are randomly oriented then we should find a relatively large fraction of very bright steep-spectrum sources with relatively weak cores, i.e., the steep-spectrum sources should greatly outnumber the flat-spectrum sources, again in contradiction with the observations (see Phinney 1985). And third, the extended radio emission in some superluminal sources (i.e., in which we are close to the jet axis) can sometimes de-project to linear sizes as large as $\sim 1\,\mathrm{Mpc}$, if the extended components and jet share a common axis, as they almost certainly do, based on comparison with more edge-on systems.

Barthel (1989) has argued that these problems disappear by postulating unification of radio-loud quasars with FR II radio galaxies (which are sometimes optically BLRGs or NLRGs); i.e., quasars whose radio axes are close to the plane of the sky are not detected as quasars, but as radio galaxies.† Barthel shows that the distribution of the angular sizes of quasars and radio galaxies is consistent with this hypothesis. Despite the large de-projected sizes of the large quasars, they are still in general smaller than the sizes of radio galaxies.

A possible problem with this unification picture is that if we compare the [O III] emission-line luminosities of radio galaxies and quasars, it is found that at a given extended radio power, the [O III] luminosity of quasars is higher by a factor of 4–10 (Baum and Heckman 1989, Jackson and Browne 1990) This should not be the case if quasars and radio galaxies are drawn from the same parent distribution and *if* the [O III] line emission is isotropic. If we are to retain this unification, we must conclude

---

† The possibility that 'unbeamed' quasars are not classified as quasars was noted by Phinney (1985).

that the [O III] emission is emitted anisotropically, preferentially along the radio axis (see Jackson and Browne 1991).

The situation is undoubtedly complicated in that luminosity dependence is probably involved in the degree of beaming and/or obscuration. For example, Lawrence (1991) notes that the narrow-line to broad-line luminosity ratio is a decreasing function of radio luminosity. In other words, the narrow lines are relatively weaker in more luminous radio sources. A physical explanation might be that the obscuring tori are thinner in higher-luminosity objects. This may be related to the problem of why no pure narrow-line quasars are seen.

It has also been suggested that lower-power (FR I) radio sources might be related to BL Lac objects (Urry, Padovani, and Stickel 1991). This picture is consistent with the known properties of the two classes, their luminosity functions and space densities – BL Lac objects have weak emission lines and show little evidence for cosmological evolution, as with FR I sources. It is clear that the parent population of the OVVs must be different from that of the BL Lacs, and indeed the likely candidate is the FR II sources (e.g., Padovani and Urry 1992). OVVs have strong emission lines and show strong evidence for cosmological evolution. It is clear moreover that BL Lacs are drawn from a different parent population than the OVVs on the basis of their soft X-ray spectra. BL Lacs, on average, have steeper X-ray spectra ($\langle \alpha_X \rangle \approx 1$) than do OVVs ($\langle \alpha_X \rangle \approx 0.5$; Worrall and Wilkes 1990). In these unification schemes, the data suggest a wide range in Lorentz factors for the jets in both FR I and FR II sources, but the number of sources at a given value of γ decreases like $N(\gamma) \propto \gamma^{-4}$ for the FR I sources, with a typical value of $\langle \gamma \rangle \approx 7.4$, and like $N(\gamma) \propto \gamma^{-2}$ for the FR II sources, yielding a typical value of $\langle \gamma \rangle \approx 11$ (Padovani and Urry 1992). These distributions of Lorentz factors are consistent with the distributions inferred from superluminal motions (Chapter 4).

7.4.3 'Grand Unification' and the Relationship Between Radio-Quiet and Radio-Loud AGNs

On balance, the evidence is quite strong for at least some unification among the various AGN phenomena. The evidence is particularly compelling for the following unification schemes:

(1) The primary difference between Seyferts and quasars is the luminosity of the central source.

(2) Seyfert 1 and Seyfert 2 galaxies are intrinsically the same sources. The difference is that in the case of Seyfert 2 galaxies we cannot see the nuclear source directly on account of our unfavorable line of sight, which is intercepted by obscuring material. A similar relationship holds between NLRGs and BLRGs.

(3) Blazars are apparently radio-loud AGNs in which our line of sight is close to the radio axis.

Is it possible to extend these to a 'grand unification' that explains the entire range of AGN phenomena? In this regard, a major question that we have not addressed is the relationship between radio-quiet and radio-loud AGNs – why do AGNs come in these two varieties? The radio-loud variety constitute a small minority of the population, except at the very high-luminosity end of the distribution, where as many as 50% or so of AGNs are radio-loud quasars such as those found in the 3C survey (Padovani 1993).

In addition to their defining differences in radio luminosity, radio-quiet and radio-loud AGNs differ in other respects. For example, Wilkes and Elvis (1987) show that the soft X-ray spectral index is flatter in radio-loud objects (i.e., $\langle \alpha_X \rangle \approx 0.5$) than in radio-quiet objects ($\langle \alpha_X \rangle \approx 1.0$).[†] The ultraviolet spectra of radio-quiet and radio-loud AGNs are virtually indistinguishable from one another (Steidel and Sargent 1991), but the optical spectra show some differences, at least statistically. Radio-quiet AGNs typically have stronger optical Fe II emission-line blends and weaker narrow-line emission (Boroson and Green 1992). Miley and Miller (1979) have found that among radio AGNs, the core-dominated flat-spectrum sources tend to have stronger optical Fe II emission blends and smoother line profiles. The lobe-dominated steep-spectrum sources have emission-line profiles that are both broader and exhibit more structure.

While radio-loud/radio-quiet unification schemes have been proposed (e.g., Scheuer and Readhead 1979), they have generally not yielded successful matches with source properties and statistics. It may indeed be that 'radio loudness' is a fundamental parameter related to the properties of the central engine, as we believe luminosity is (Chapter 3). For example, Blandford (1990) has attempted a qualitative phenomenological classification based on three fundamental physical parameters: (a) the mass of the central engine, (b) the accretion rate (relative to the Eddington rate), and (c) the angular momentum of the central source, plus the inclination of the system as seen by the observer. The importance of angular momentum is that the orbital motions of charged particles around the black hole determine whether or not electromagnetic effects will be powerful enough to generate and collimate relativistic jets.

Wilson and Colbert (1995) have considered the possibility that the fundamental parameter that determines whether a source is radio quiet or radio loud is the black-hole spin rate. The assumption in their model is that the X-ray through IR spectrum is thermal in origin, driven by mass accretion onto a black hole. The radio energy (which appears in the form of jets) is mechanical energy extracted from the black hole. The key element in producing the bimodal radio-luminosity distribution is the assumption that the black-hole spin rates are determined not by the normal accretion process but by mergers of large black holes following the mergers of their parent galaxies. The outcome of the merger depends on the relative sizes of the initial black holes, which might plausibly be related to the masses of the parent galaxies (so thus the most massive black holes are relatively rare). High spin rates are achieved by mergers

[†] See also Fig. 4.1.

of black holes of comparable mass. Mergers of pairs of galaxies can thus yield the following:

(1) A pair of low-mass (say $10^5 M_\odot$) objects yields a low-mass black hole with a high spin rate. This might produce the low-luminosity parsec-scale radio sources seen in early-type galaxies.

(2) A high-mass (say $10^8 M_\odot$)/low-mass pair will produce a high-mass black hole with low spin, presumably yielding a radio-quiet AGN.

(3) The rarest of all mergers would be two high-mass objects, yielding a high-mass, high-spin black hole that would produce a luminous radio-loud AGN.

The basic model has a number of attractive features that appear to account for many of the observations.

8 The Environment of AGNs

As we pointed out in Chapter 3, the main problem with sustaining an active nucleus by gravitational accretion over its lifetime of at least 10^8 years is funneling enough mass into the nucleus. Removing a sufficient amount of angular momentum from the gas flowing into the nucleus requires breaking the azimuthal symmetry of the galaxy's gravitational potential. A clear way to do that is by gravitational interactions with other systems, as was originally suggested by Toomre and Toomre (1972) and by Gunn (1979). This provides motivation for examining the nearby environment of AGNs to see if indeed there is evidence for interactions with nearby galaxies. The two specific questions that we want to consider are:

(1) What kinds of galaxies harbor AGNs? Are there any discernible differences between galaxies with active nuclei and those without them?
(2) Does the presence or absence of companion galaxies have anything to do with whether or not a galaxy harbors an AGN?

We will consider these issues separately, although they are clearly related.

8.1 Host Galaxies

The study of the 'host galaxies', those galaxies that contain active nuclei, is a very difficult undertaking. The major problems were alluded to at the beginning of this book: the light from the AGN itself often dominates the total light from the galaxy, particularly in the case of the highest-luminosity AGNs, which are spatially rare and thus typically found only at great distances. Consequently the work on the lower-luminosity end of the AGN distribution, i.e., Seyfert galaxies, has tended to yield less ambiguous results.

8.1.1 Host-Galaxy Morphology

Seyfert host galaxies have been the focus of several investigations whose goal is to determine whether or not the galaxies that harbor active nuclei are different in any other sense from galaxies which do not have active nuclei, or if different types of galaxies share some common characteristics that may or may not be correlated with the presence of nuclear activity. The first important morphological study of Seyfert galaxies by Adams (1977) revealed that most Seyfert galaxies are spiral galaxies. While the original

Table 8.1

**Percentages of Galaxies with
Morphological Irregularities**[a]

Feature	Seyfert Galaxies	Non-Seyfert Galaxies
Inner Rings	57%	35%
Outer Rings	43%	8%
Both Types	30%	3%

[a] Simkin, Su, and Schwarz (1980).

galaxies in Seyfert's (1943) study were all spirals (as pointed out by van den Bergh 1975), the definition of 'Seyfert galaxies' had gradually become spectroscopic, so this was not necessarily a foregone conclusion.† A later study by Heckman (1978) showed that in fact Seyfert host galaxies tend to be, but are not exclusively, earlier-type spirals (i.e., Sa–Sb). A survey of well-resolved Seyferts (i.e., nearby, with $cz < 5000 \, \mathrm{km \, s^{-1}}$, or within $50h_0^{-1}$ Mpc) by Simkin, Su, and Schwarz (1980) revealed that Seyfert host galaxies often show signs of morphological irregularities of the type which are often attributed to tidal interactions. The frequencies with which these irregularities are seen are summarized in Table 8.1. The preponderance of disturbed morphologies among Seyfert host galaxies has been confirmed by MacKenty (1990) and others. Radio observations of H I 21-cm emission from Seyfert galaxies also reveal morphological or line-profile peculiarities compared to otherwise similar normal galaxies (Heckman, Balick, and Sullivan 1978, Mirabel and Wilson 1984). In addition to these features, more than half of Seyfert host galaxies are found to have barred structure, although only about one third of normal (i.e., those without active nuclei) galaxies show bars at the same spatial resolution. However, we must keep in mind that not all barred spirals contain AGNs, and there are many AGNs which are in galaxies with no discernible bar. Furthermore, a more recent study (McLeod and Rieke 1995a) indicates that the evidence that Seyfert nuclei occur preferentially in barred systems is marginal at best.

QSO host galaxies are by definition much more difficult to study as the starlight component is hard to detect in the glare of the active nucleus, although it has long been clear that QSOs are not truly point-like sources (Matthews and Sandage 1963). An extensive study of low-redshift QSO host galaxies by Smith *et al.* (1986) indicates that, as with Seyfert galaxies, approximately 50% of the host galaxies show morphological peculiarities.

We caution that the statistics upon which all of these conclusions have been reached

† Recall from Chapter 1, however, that an elliptical galaxy with a bright nucleus was likely to be classified as an 'N' galaxy rather than as a Seyfert galaxy. The early classification of active galaxies was thus highly uncertain, and many biases towards specific properties (such as spiral hosts) are likely to remain in any list of known Seyfert galaxies.

are based on samples limited in number and in spatial resolution. We expect that the statistics will improve as additional studies are undertaken with both *HST* and ground-based telescopes.

8.1.2   Surface-Brightness Profiles and Luminosities

Much of the work on host galaxies in both Seyferts and QSOs has been directed towards study of their surface-brightness profiles. The surface-brightness profiles $\Sigma(r)$ (in units of flux per unit area, i.e., ergs $s^{-1}\,cm^{-2}\,arcsec^{-2}$) of normal galaxies can generally be described by

$$\log \frac{\Sigma(r)}{\Sigma_0} = -\left(\frac{r}{r_s}\right)^{1/\beta},$$

(8.1)

where Σ_0 is the central surface brightness and r_s is a scale length (Mihalas and Binney 1981). Disk systems, such as spirals and S0s, are characterized by $\beta = 1$, i.e., the surface brightness of the galaxy decreases exponentially with scale length r_s (Freeman 1970). Spheroidal systems, which include not only elliptical galaxies but the central bulges of spirals as well, are characterized by $\beta = 4$ (de Vaucouleurs 1953).† It should be emphasized that eq. (8.1) is an *empirical* fit to the data for both disk and spheroidal systems; there is no known physical reason why the surface-brightness profiles should have this particular form. We also note that the actual fits to the data usually assume that the systems are intrinsically thin circular disks, and that the ratio of the apparent sizes of the minor and major axes b/a is equal to $\cos i$, where i is the inclination of the galaxy. The surface-brightness fits are then usually done by azimuthally averaging the data. Features like spiral arms, rings, and bars are generally of low enough contrast that they do not greatly affect the final results.

It is usually assumed that the surface-brightness profiles of AGNs can be modeled in the same way as normal galaxies, but with an additional central point source that represents the AGN continuum source.‡ The observed surface-brightness profile of the AGN continuum source is thus described by the 'point-spread function' (PSF) for the telescope and detector being used. For ground-based telescopes, the PSF width is determined primarily by atmospheric seeing. In real systems, the PSF can usually be modeled with a Gaussian core whose width is primarily seeing dependent, plus extended wings, whose strength and extent depend on the optical system. In space-based astronomy, the PSF is determined entirely by the optical system, and in complex

---

† The $\beta = 4$ case is more conventionally written in the form

$$\frac{\Sigma(r)}{\Sigma_e} = \mathrm{dex}\left\{-3.33\left[(r/r_e)^{1/4} - 1\right]\right\},$$

where r_e is the 'effective radius' which contains half of the total light, and $\Sigma_e = \Sigma(r_e)$. This is known as the 'de Vaucouleurs $r^{1/4}$ law'. When expressed in the form of eq. (8.1), $r_e = (3.33)^4 r_s$ and $\Sigma_0 = 10^{3.33} \Sigma_e$.

‡ The BLR is also effectively a point source and cannot be separated from the continuum source by standard imaging techniques. While the NLR is somewhat extended, it usually can be ignored as its total contribution to the broad-band light is very small because the emission is confined to a small range in wavelength.

systems, such as the imaging cameras on the refurbished *HST*, the PSF may have gross deviations from azimuthal symmetry.

Determination of the surface-brightness profiles for AGNs thus involves a multiple-component fit to the data. The parameters we seek are the amplitude of the central (PSF-like) component, the host-galaxy central surface brightness Σ_0, and the scale length r_s. Depending on how well the host galaxy is resolved, (a) the galaxy may be modeled with two components, a disk plus a bulge, (b) the value of β may be fixed if the morphology of the host galaxy is apparent, or (c) fits using both $\beta = 1$ and $\beta = 4$ can be attempted. The last strategy is commonly used in the study of QSO host galaxies at redshifts and levels of photometric contrast where the galaxy structure is not easily determined. On the basis of the fitted parameters, the total flux from the host galaxy can be estimated by integrating eq. (8.1), i.e.,

$$F_{\text{tot}} = 2\pi \int_0^\infty \Sigma(r)\, r\, dr = \frac{(2\beta)!}{(2\ln 10)^{2\beta}}\, \pi\Sigma_0\, r_s^2. \qquad (8.2)$$

While this procedure is straightforward in principle, in practice it is fraught with many difficulties. In addition to the obvious problems of limited angular resolution and sky background, the imaging system employed in such investigations must be characterized by a large dynamic range (i.e., ability to record a wide range of surface brightness) and linear response; consequently charge-coupled devices (CCDs) have supplanted virtually all other detectors for this type of work. We note, however, that it does not necessarily follow that space-based observations are always superior to ground-based observations. The much narrower PSF for a good space-based camera places a severe constraint on the dynamic range required, since the contrast between the brightness of the center of the AGN and the outer regions of galaxy is greater for a narrower PSF.

8.1.3 Relationships Between AGNs and Their Hosts

There is a wide consensus based on many different imaging studies that radio-quiet QSOs and Seyfert galaxies tend to be found in disk systems and radio-loud QSOs and BLRGs tend to be found in elliptical galaxies (e.g., Gehren *et al.* 1984, Malkan 1984, Smith *et al.* 1986, Hutchings, Janson, and Neff 1989). This is not a completely clear-cut dichotomy, but it seems to be true in general. Much of its general acceptance is probably due to the fact that the most luminous radio galaxies are ellipticals whereas spirals tend to be radio quiet. The colors of the host galaxies are generally found to be consistent with their inferred morphological type, but Kotilainen and Ward (1994) note that the host galaxies of Seyfert 1 nuclei tend to be somewhat redder than normal galaxies of the same luminosity and morphology, possibly because of emission from dust heated by the AGN.

Another result common to numerous studies is that there is a rather good correlation between the absolute magnitudes of the host galaxies and the AGNs they harbor, i.e., brighter AGNs are found in more luminous galaxies (Fig. 8.1). While the process

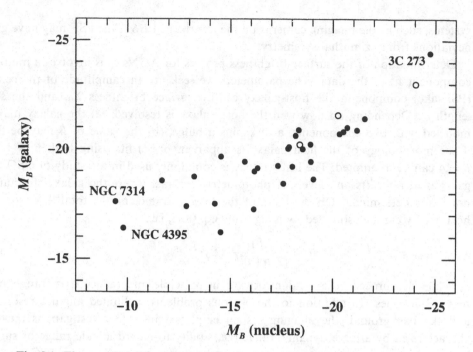

Fig. 8.1. The correlation between the absolute B magnitudes of the host galaxies and the AGNs they harbor. Notably bright and faint sources are indicated. The open circles refer to host galaxies of BL Lac objects. Luminosities are calculated assuming $h_0 = 1$ and $q_0 = 0.5$ (see Chapter 9). Adapted from Kotilainen and Ward (1994).

of fitting surface-brightness profiles yields fairly significant uncertainties for both Σ_0 and r_s, the product $\Sigma_0 r_s^2$ is generally much better determined (cf. eq. 8.2), so that the total luminosity of the host galaxy is measured with reasonable accuracy (Malkan, Margon, and Chanan 1984, Véron-Cetty and Woltjer 1990). We note, however, that it is also frequently pointed out that there is a potentially important selection effect that can influence this result, namely that the lower right part of the diagram (bright AGNs in faint galaxies) may be empty because the stellar envelopes are too faint to be detected in this case, and the upper left (faint nuclei in bright galaxies) may be selected against as these are Seyferts, not QSOs. Indeed, such low-luminosity galaxies, 'dwarf Seyferts' (Ho, Filippenko, and Sargent 1995), appear to be quite common. In a spectroscopic survey of the nuclei of a large number of relatively nearby spiral galaxies, Ho, Filippenko, and Sargent (1994) find low-luminosity Seyfert nuclei or LINER nuclei in $\sim 40\%$ of these galaxies, most of which had not previously been known to harbor active nuclei.

Smith *et al.* (1986) argue that QSOs are found only in the most luminous galaxies, as would be expected from Fig. 8.1. They also conclude that the host galaxies of radio-loud QSOs are likely to be around twice as luminous as the host galaxies of radio-quiet QSOs, which is also in agreement with other studies (Malkan 1984, Hutchings 1987,

Table 8.2

Absolute *B* Magnitudes of Types of Host Galaxies

Type of Object	M_B (mag)
All QSO Host Galaxies[a]	$-20.7 + 5 \log h_0$
Host Galaxies of Radio-Loud Quasars[a]	$-21.1 + 5 \log h_0$
Host Galaxies of Radio-Quiet QSOs[a]	$-20.4 + 5 \log h_0$
Host Galaxies of BL Lac Objects[b]	$-21.5 + 5 \log h_0$
Seyfert Host Galaxies[a]	$-19.4 + 5 \log h_0$
Fiducial Bright Normal Galaxy[c]	$-19.1 + 5 \log h_0$

[a] Smith *et al.* (1986).

[b] Ulrich (1989).

[c] Luminosity L^* in the Schechter (1976) luminosity function (Chapter 11).

Véron-Cetty and Woltjer 1990). Typical absolute *B* magnitudes† of representative samples of various objects are given in Table 8.2.

Recent *HST* imaging results on QSOs have been challenging conventional views about QSO host galaxies (Bahcall, Kirhakos, and Schneider 1995). Even though the number of QSOs that have been studied in detail is still limited, the new data strongly indicate that the correlation between nuclear and host-galaxy luminosity is a good deal worse than implied by Fig. 8.1, in particular because of failure to detect any host galaxy at all in several cases. However, McLeod and Rieke (1995b) argue that the original analysis was not sensitive to smooth, structureless galaxies, and that the luminosities and optical–infrared colors are consistent with early-type luminous host galaxies in all cases. Clearly additional observations must be undertaken and possible difficulties with the challenging problem of deconvolution of *HST* images need to be addressed more completely before the standard view is rejected.

Spectroscopic studies of the faint extended emission around QSOs yield results consistent with the imaging studies (e.g., Boroson, Oke, and Green 1982). Figure 8.2 shows a spectrum of the extended emission around 3C 48. This spectrum is of particular historical significance as it was the first irrefutable direct evidence that QSOs reside in stellar systems (Boroson and Oke 1982). These spectra often show weak nebular emission lines in addition to the continuum, but the absorption features that arise in stars are often weak and hard to detect in these off-nuclear spectra. In some cases, the colors and spectral features indicate the presence of a young population of stars (e.g., Boroson and Oke 1982, Hutchings and Crampton 1990).

It is also worth comparing the optical luminosities of the host galaxies of radio-loud

† The astute reader will recognize that a *B*-band magnitude of a redshifted object will be different from that of an identical zero-redshift (local) object simply because a different part of the spectrum is observed. This requires a redshift-dependent adjustment called a *K*-correction, which is discussed in Chapter 10.

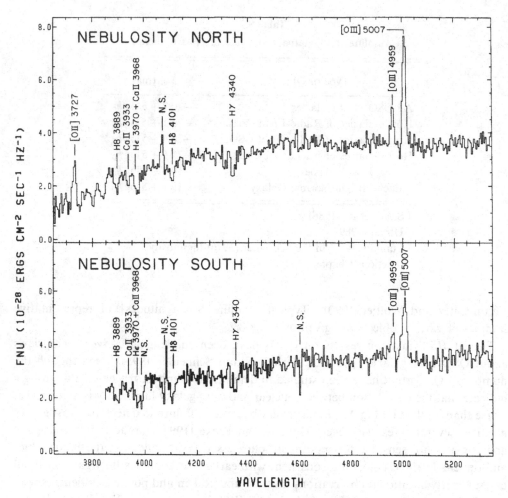

Fig. 8.2. The spectrum of the faint nebulosity immediately north and south of the quasar 3C 48 ($z = 0.367$), obtained by careful subtraction of the nuclear (AGN) light from the low surface-brightness extended emission (Boroson and Oke 1982). In addition to weak, narrow emission lines, absorption lines characteristic of a young stellar population (integrated spectral type approximately A7) are also detected. This observation constituted the first direct proof that high-luminosity quasars reside in galaxies. Features labeled 'N.S.' are artifacts of incomplete subtraction of strong night-sky emission lines. Figure courtesy of T. A. Boroson. Reproduced by permission from *Nature*, Vol. 267, pp. 397–399. Copyright 1982 Macmillan Journals Limited.

quasars with those of FR II sources that do not show optical evidence of activity, since in some unified schemes these should be the same objects (Barthel 1989; see Chapter 7). Véron-Cetty and Woltjer (1990) conclude that these two types of galaxies are statistically indistinguishable. Other studies, however, have found that the host galaxies of radio-loud QSOs are more luminous than optically inactive radio galaxies, typically by ~ 0.5–1.0 mag for similar galaxies of comparable radio luminosity (Hutchings 1987, Smith and Heckman 1989).

8.2 Nearby Companions

In principle, the large-scale (Mpc) environment of AGNs might provide important clues as to the kind of physical conditions that lead to AGN activity. Quantifying the possible relationship between AGNs and their neighbors is not a simple undertaking; for example, we have to guard against the possibility of interpreting chance alignments as 'interacting systems'. More subtle interactions may not be readily detectable as morphological irregularities, but might be hinted at statistically in terms of the space density of galaxies around AGNs. In drawing statistical inferences about the frequency with which we find companion galaxies near AGNs, we must carefully select suitable 'control' fields around otherwise similar normal galaxies. Moreover, we must exercise some care in defining what we consider to be an interacting system: in many published studies, systems are defined to be interacting if they meet certain semi-quantitative criteria such as overlapping outer isophotes and near coincidence in redshift (i.e., usually within radial-velocity differences $\Delta v \lesssim 10^3$ km s^{-1}, which is about the maximum radial velocity difference expected between two cluster members).

There have been many attempts to study the immediate environments of both Seyferts and quasars. A clear consensus that has emerged is that AGNs are more likely than other galaxies to have companion systems.† Dahari (1984) found that $\sim 15\%$ of Seyfert galaxies have companions (meeting well-defined criteria), whereas only $\sim 3\%$ of a control sample of normal galaxies have such companions. MacKenty (1990) reaches a similar conclusion, based on a larger, better-defined sample. Kollatschny and Fricke (1989) find that companions of Seyfert galaxies are also likely, relative to companions of normal galaxies, to show strong emission lines in their spectra.

In the case of QSOs, it has been known at least since the pioneering work of Kristian (1973) and Stockton (1978)‡ that the fields of QSOs contain a surprisingly large number of faint galaxies close to the same redshift as the QSO. Weymann *et al.* (1978) showed that Stockton's data indicate that the probability of finding a galaxy at any distance r from a chosen galaxy is higher if the reference galaxy contains an AGN. Similar conclusions about the tendency of AGNs to be found in groups or clusters have been reached in other investigations (e.g., Heckman *et al.* 1984, Yee and Green 1984, Hutchings, Crampton, and Campbell 1984). The companion galaxies themselves do not appear to be in any way affected by the presence of the AGN, as they do not seem to be unusually luminous (Yee 1987); conclusions about their colors are less definitive.

---

† The term 'companion' is generally taken to mean a galaxy which is close enough that the systems may interact gravitationally, although it does not necessarily mean that the systems are gravitationally bound.

‡ We single out Stockton (1978) as of particular significance in our opinion for demonstrating definitively that QSOs are at the cosmological distances implied by their large redshifts.

A related topic, first considered by Osmer (1981), is that of QSO clustering – are QSOs themselves distributed at random in space, or do they have a tendency to cluster as galaxies do? Recent work in this area (Shanks and Boyle 1994) suggests that the clustering characteristics of QSOs are no different that those of non-active galaxies, which *do* show clustering on scale lengths less than $\sim 10h_0^{-1}$ Mpc.

While AGNs are frequently found with companions and are rarely isolated systems, they are conspicuously absent in rich clusters (Dressler, Thompson, and Shectman 1985). This statement is quite firm for radio-quiet QSOs, but luminous radio-loud quasars tend to be found in richer environments, and this tendency seems to increase dramatically for redshifts larger than $z \approx 0.3$ (Yee and Green 1987, Ellingson, Yee, and Green 1991).

The high incidence of morphological peculiarities and of close companions is indeed suggestive that interactions with other systems *can* plausibly trigger AGN activity. However, neither close companions nor disturbed morphology provide either necessary or sufficient conditions for producing AGNs. For example, strongly interacting systems (such as those studied by Arp) tend *not* to show AGN phenomena, and some AGNs are quite isolated with no obvious companion systems.

8.3　Galaxy Mergers and Starbursts

As noted in Chapter 7, mergers of massive black holes might be the process that triggers the AGN phenomenon. Mergers of galaxies are an especially effective way for both stars (Roos 1985) and gas (Barnes and Hernquist 1991) to lose angular momentum and be carried in towards the nuclear regions. Signs of a recent merger should be detectable in such galaxies. Indeed, there are some AGNs and AGN-like objects that have multiple nuclei (e.g, Petrosian, Saakian, and Khachikian 1979, Kollatschny, Netzer, and Fricke 1986), which might be the signature of a recent merger, although in any individual case the possibility that an apparent multiple-nucleus source is chance projections of physically unrelated sources cannot be easily dismissed. About 5% of the Markarian galaxies show evidence for multiple nuclei. These systems are identified morphologically (i.e., spatially resolved nuclei) and/or spectroscopically (i.e., multiple-redshift systems in spectra), and generally contain either a Seyfert nucleus plus a starburst, or more frequently, two starburst nuclei. These systems often show the kinds of morphological irregularities attributed to mergers of galaxies, such as tidal tails.

There is some evidence that AGNs are associated with starbursts. Indeed, there have been numerous suggestions in the literature that they are somehow manifestations of the same basic phenomenon, or that they are related in an evolutionary sense. For example, Weedman (1983) and Norman and Scoville (1988) have argued that ultimately starbursts lead to the formation of black holes. Conversely, it has been argued that the presence of an AGN in a galaxy might trigger a burst of star formation (Rees 1989, Daly 1990). At this time, however, the evidence for either case remains fragmentary.

9 The Geometry of the Expanding Universe

Because the highest-luminosity AGNs can be detected at very large distances (and hence 'lookback' times) they provide an important probe of the history of the Universe. Luminous quasars are detected at redshifts up to $z \approx 5$; we see these objects as they were when the Universe was $\sim 10\%$ its current age. Quasars are the most distant discrete objects that we have observed, and they are therefore of tremendous importance in understanding the formation of discrete structures from the primordial gas and as measures of the first appearance of metals. Furthermore, as we shall see in subsequent chapters, the luminosity function of quasars has varied over the observable history of the Universe, and this has something to tell us about the formation and evolution of galaxies and photoionization equilibrium of the intergalactic medium as a function of redshift, among other things. Finally, as we will see in Chapter 12, quasars can be used as background sources against which we can detect the absorption signatures of less luminous objects, thus providing us with a probe of otherwise unobservable gas at high redshifts.

Determination of the space density and luminosity function of any type of astronomical object is difficult, primarily because of the different volumes over which the most luminous and least luminous objects of a given class can be detected. In the case of quasars, the situation is further complicated by the fact that quasars are observed at sufficiently large distances that the curvature and expansion of the Universe must be taken into account. In this chapter, we introduce some of the concepts and develop some of the mathematical formalism that we will need in order to study the luminosity and space distribution of quasars. The notation used in this chapter is largely consistent with that used by Peebles (1971, 1993), and much of the discussion in this chapter can be traced to these sources and to Weinberg (1972).

9.1 The Metric

The first problem we must confront is how to measure distances and time in a four-dimensional space-time continuum. Consider first how we do this in a three-dimensional Euclidean space. In Euclidean space, we can establish an infinitesimal line element

$$d\tilde{s}^2 = dx^2 + dy^2 + dz^2. \tag{9.1}$$

This line element is invariant under coordinate transformation; i.e., its magnitude is the same regardless of the coordinate system we choose.

135

Extension of this principle to additional dimensions is straightforward. However, the fourth dimension in this case (time) needs special treatment, as the fundamental postulate of relativity is that it is the *speed of light* that is the invariant for all observers. The 'metric', or line element, we use in this case is the Minkowski metric of special relativity,

$$ds^2 = c^2dt^2 - dx^2 - dy^2 - dz^2 = c^2dt^2 - d\vec{s}^2, \tag{9.2}$$

which is invariant under coordinate transformations. It is sometimes more useful to write this in spherical coordinates as

$$ds^2 = c^2dt^2 - dr^2 - r^2\left(d\theta^2 + \sin^2\theta d\phi^2\right). \tag{9.3}$$

The paths of freely moving objects through space-time follow the shortest possible paths between two points. For photons, such paths are known as 'null geodesics' or 'great circles', and these are described by

$$ds^2 = 0. \tag{9.4}$$

The Minkowski metric applies in the very special case of Euclidean space (hence the name 'special relativity'). We seek a similar metric that will apply to the more general case of a non-Euclidean geometry. However, we can restrict consideration of possible geometries by making some basic assumptions about the nature of the Universe. Specifically, we make the following assumptions:

Homogeneity: It is assumed that the gross properties of the Universe are the same everywhere, if one averages over a suitably large volume. In actual practice, this means on distance scales of ~ 100 Mpc.

Isotropy: It is assumed that the Universe looks approximately the same in each direction *for every observer*.

For our particular location in the Universe, isotropy is fairly well established – to first approximation, the distribution of galaxies is about the same in every direction (accounting for obscuration due to dust in our own Galaxy and averaging over structures such as clusters of galaxies), and the cosmic background radiation from the big bang shows isotropy to a very high degree (Smoot *et al.* 1992). However, that isotropy holds for all observers at all locations is an assumption, although careful consideration of the cosmic background radiation indicates that it is probably a pretty reasonable assumption (Goodman 1995). It can be shown that if isotropy holds everywhere, homogeneity also obtains.

The assumption of homogeneity and isotropy is equivalent to the statement that there is nothing special about our particular location in the Universe; this is known as the 'cosmological principle'. The philosophical underpinnings of the cosmological principle are essentially Copernican. Our early assumptions that the Earth was the center of the Solar System turned out to be mistaken. Similarly, early star-count analyses by Herschel and Kapteyn pointed towards a Sun-centered Universe, and only Shapley's studies of globular clusters and Trumpler's recognition of the effects of interstellar absorption by dust corrected this second error. The cosmological principle is an attempt to avoid yet another error of the same type.†

On the basis of the assumptions of homogeneity and isotropy, a more general form of the metric was derived independently by Robertson (1935) and Walker (1936) for a multi-dimensional geometry. The 'Robertson–Walker metric' is

$$ds^2 = c^2dt^2 - \frac{a(t)^2dr^2}{(1 - kr^2/R^2)} - a(t)^2r^2\left(d\theta^2 + \sin^2\theta d\phi^2\right). \tag{9.5}$$

In this equation, r, θ, and ϕ are coordinates chosen by a particular observer, and these form a grid which can expand and contract as the size of the Universe changes (clusters of galaxies, for example, have fixed coordinates r, θ, and ϕ). These fixed coordinates expand and contract with the evolving Universe, and are thus referred to as 'comoving coordinates'. The 'expansion parameter' $a(t)$ reflects the physical scale of the Universe as it varies with time; by including the parameter a, one can obtain the proper distances between comoving coordinates as a function of time. The parameter R is a scaling constant, which in some geometries can be interpreted as the radius of curvature of the Universe at the present epoch. The constant k reflects the *sense* of the curvature of space, and has values ± 1 or 0 only.‡ Proper time is the time measured by a clock in the observer's reference frame; every observer at a fixed comoving coordinate in the Universe measures the same proper time. By convention, we define 'cosmic time' t as the proper time measured from the big bang. The present cosmic time will be denoted as t_0; similarly, the values of all variable parameters at the present epoch will be denoted with the same subscript zero, e.g., H_0, q_0. Also, in this convention, $a(t_0) \equiv a_0 = 1$, which follows automatically from our adopting the constant R as the *current* radius of curvature.

Note that in the special case $k = 0$, eq. (9.5) reduces to a form distinguished from the Minkowski metric (eq. 9.3) by the presence of the expansion parameter a, which allows the comoving coordinate grid to expand or contract with time.

9.1.1 The Cosmological Redshift

For convenience and without loss of generality, we can consider ourselves to be at the origin $r = 0$. Suppose that at some cosmic time t_1 a galaxy at coordinates (r_1, θ_1, ϕ_1) emits a photon in the $-r$ direction, which arrives here at current time t_0. Its path is

---

† We note in passing that in some cases a *third* assumption has been made, namely that the Universe is 'not only homogeneous and isotropic, but also that its gross properties do not change with time. This was sometimes known as the 'perfect cosmological principle' and was the basis of the steady-state theories of Hoyle and others. By and large, steady-state theories have failed to account for the cosmic background radiation, in particular, and we will not consider them here.

‡ In some notations, the constant k is absorbed into R^2. Thus, in the case $k = -1$, the 'radius' of the Universe R is imaginary.

given by eq. (9.4) and eq. (9.5),

$$c\,dt = \frac{a(t)\,dr}{\left(1 - kr^2/R^2\right)^{1/2}}. \tag{9.6}$$

Moving the time-dependent parameters to the left side of the equation and integrating over the path of the photon we get

$$c\int_{t_1}^{t_0} \frac{dt}{a(t)} = \int_0^{r_1} \frac{dr}{\left(1 - kr^2/R^2\right)^{1/2}} \equiv f(r_1), \tag{9.7}$$

which is independent of time. The function $f(r_1)$ can be evaluated for the possible values of k:

$$\begin{aligned}
f(r_1) &= \int_0^{r_1} \frac{dr}{\left(1 - r^2/R^2\right)^{1/2}} = R\sin^{-1}\frac{r_1}{R} \qquad (k = +1)\\[2mm]
&= \int_0^{r_1} dr = r_1 \qquad\qquad\qquad\quad (k = 0)\\[2mm]
&= \int_0^{r_1} \frac{dr}{\left(1 + r^2/R^2\right)^{1/2}} = R\sinh^{-1}\frac{r_1}{R} \qquad (k = -1).
\end{aligned} \tag{9.8}$$

Note that in each case, in the limit of $r_1 \ll R$, $f(r_1) \approx r_1$.

Imagine now two successive wavefronts emitted by the galaxy, the first at t_1 and the second at $t_1 + \delta t_1$. These are received by us at the origin at the corresponding times t_0 and $t_0 + \delta t_0$. The path for the second front is similarly given by

$$c\int_{t_1+\delta t_1}^{t_0+\delta t_0} \frac{dt}{a(t)} = \int_0^{r_1} \frac{dr}{\left(1 - kr^2/R^2\right)^{1/2}} = f(r_1). \tag{9.9}$$

Over a single wave oscillation ($\sim 10^{-14}$ s for visible light), $a(t)$ can be taken to be constant, so the difference in the two integrals above is the difference in the areas of the two strips in Fig. 9.1, which must be equal to zero, i.e.,

$$\frac{\delta t_1}{a(t_1)} = \frac{\delta t_0}{a(t_0)}. \tag{9.10}$$

Thus,

$$\frac{\delta t_1}{\delta t_0} = \frac{a(t_1)}{a(t_0)}. \tag{9.11}$$

Since the wave period $\delta t \propto 1/v = \lambda/c$, we have

$$\frac{\lambda_1}{\lambda_0} = \frac{a(t_1)}{a(t_0)}. \tag{9.12}$$

By definition, the redshift z is given by

$$z = \frac{\lambda_0 - \lambda_1}{\lambda_1}, \tag{9.13}$$

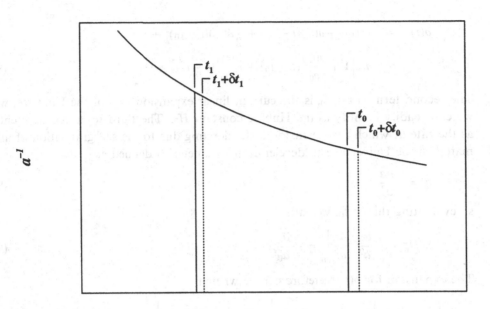

Time

Fig. 9.1. As shown in eqs. (9.7) and (9.9), the integral of $dt/a(t)$ for two photons traveling from r_1 to the observer at successive departure times t_0 and $t_0 + \delta t_0$ and corresponding arrival times t_1 and $t_1 + \delta t_1$ must be the same. The difference between the two integrals is the difference between the areas of the two approximately rectangular areas of respective areas $\delta t_1/a(t_1)$ and $\delta t_0/a(t_0)$; for the difference between the integral to vanish, these areas must be identical (eq. 9.10).

so this becomes

$$1 + z = \frac{a_0}{a(t_1)}. \tag{9.14}$$

This shows that photons emitted at a cosmic time t_1 which are arriving here at the present epoch are *redshifted* by a factor proportional to the ratio of the size of the Universe at the current epoch and at time t_1. On the basis of the Robertson–Walker metric only, we have shown that the cosmological redshift z results from the changing size of the Universe.

9.1.2 The Expansion Parameter

Given no *a priori* knowledge of how the expansion parameter varies with time, we can attempt to describe its behavior at some time t not too different from the present through a Taylor expansion around its current value, i.e.,

$$a(t) = a_0(t_0) + \dot{a}(t_0)(t - t_0) + \frac{1}{2}\ddot{a}(t_0)(t - t_0)^2 + \cdots$$

$$= a_0\left[1 + \frac{\dot{a}(t_0)}{a_0}(t - t_0) + \frac{\ddot{a}(t_0)}{2a_0}(t - t_0)^2 + \cdots\right]. \tag{9.15}$$

The second term, $\dot{a}(t_0)/a_0$, is the current linear expansion rate of the Universe, which we can readily identify as the Hubble constant H_0. The third term can be identified as the rate at which the expansion is decelerating due to the self-gravitation of all the matter in the Universe. The 'deceleration parameter' is defined as

$$q = -\frac{\ddot{a}a}{\dot{a}^2}, \tag{9.16}$$

so evaluating this at t_0, we have

$$q_0 H_0^2 = -\frac{\ddot{a}a}{\dot{a}^2}\frac{\dot{a}^2}{a^2}\bigg|_{t_0} = -\frac{\ddot{a}_0}{a_0}. \tag{9.17}$$

The expansion for $a(t)$ therefore can be written

$$a(t) = a_0\left[1 + H_0(t - t_0) - \frac{1}{2}q_0 H_0^2(t - t_0)^2 + \cdots\right]. \tag{9.18}$$

From the definitions of H_0 and q_0, we see that the units of H_0 must be time^{-1} and that q_0 must be dimensionless. Note that if $q_0 = 0$, the third term vanishes and the expansion is linear, in the absence of higher-order terms. If $q_0 > 0$, the effect of the third term is to decelerate the expansion.

9.1.3   Distance Measures

Consider first the 'proper distance' d_p, which is the real physical distance as measured with fixed rods between two points (say galaxies) at some fixed cosmic time. We can obtain this distance by integrating over the Robertson–Walker metric with t fixed, i.e.,

$$d_p = \int_{t \text{ (fixed)}} ds = \int_0^{r_1} \frac{a(t)dr}{(1 - kr^2/R^2)^{1/2}} = a(t_0)f(r_1). \tag{9.19}$$

The time rate of change of the proper distance is

$$\dot{d}_p = \dot{a}(t_0)f(r_1), \tag{9.20}$$

and using eq. (9.19) this can be written

$$\dot{d}_p = \dot{a}(t_0)\left(\frac{d_p}{a_0}\right) = H_0 d_p, \tag{9.21}$$

which is the Hubble law.

Consider now an object of linear size D which subtends an angle $\delta\theta$ on the sky. In Euclidean geometry, we define a distance

$$d_A = \frac{D}{\delta\theta}. \tag{9.22}$$

We can use the Robertson–Walker metric to show that the size of this object at coordinate r_1 is

$$ds = a(t_1)r_1\delta\theta = D \tag{9.23}$$

so that we can define an 'angular size distance'

$$d_A = \frac{D}{\delta\theta} = a(t_1)r_1. \tag{9.24}$$

We note that

$$\frac{d_p}{d_A} = \frac{a_0(t_0)f(r_1)}{a(t_1)r_1} = (1+z)\frac{f(r_1)}{r_1}. \tag{9.25}$$

For $r_1 \ll R$, this is approximately equal to $1+z$.

So far, all of these results are simply geometrical results that depend only on the assumptions of isotropy and homogeneity. Note in particular that we have not yet included any dynamics at all.

9.2 The Friedmann–Lemaître Equations

By combining the Robertson–Walker metric with the Einstein field equations for general relativity, it is possible to derive two equations of motion for the Universe. These are known as the Friedmann–Lemaître equations (e.g., Lemaître 1931). While a precise derivation of these equations is beyond the scope of this book, we can get the general flavor of the physics involved by carrying out an oversimplified Newtonian derivation. We begin by noting that for $r \ll R$, the proper distance ℓ to a particular galaxy can be approximated as

$$\ell = a(t)r, \tag{9.26}$$

where r is the fixed, comoving coordinate for the galaxy.

One of the basic results of general relativity is that we can treat any sub-volume of space as though it were isolated and consider only the effects of mass within the sub-volume. This is a consequence of a more general result known as Birkhoff's theorem, and it is analogous to Newton's demonstration that for a spherically symmetric mass distribution, the acceleration on a test particle depends only on the mass enclosed by a sphere centered on the origin on whose surface the test particle is located. Furthermore, on small enough spatial scales, curvature can be neglected and space can be considered to be locally flat or Euclidean. Thus, the equation of motion for our isolated sub-volume can be written as

$$\frac{d^2\ell}{dt^2} = -\frac{GM}{\ell^2} = \ddot{\ell}, \tag{9.27}$$

where M is the mass within the sub-volume. By multiplying this by $\dot{\ell}$ we get

$$\dot{\ell}\ddot{\ell} = -\frac{GM\dot{\ell}}{\ell^2}. \tag{9.28}$$

Here we note that $d\dot{\ell}^2/dt = 2\dot{\ell}\ddot{\ell}$ and that $d(\ell^{-1})/dt = -\dot{\ell}/\ell^2$, so we can rewrite eq. (9.28) as

$$\frac{d}{dt}(\dot{\ell}^2) = \frac{d}{dt}\left(\frac{2GM}{\ell}\right).$$

(9.29)

Integrating this gives

$$\dot{\ell}^2 = \frac{2GM}{\ell} + C,$$

(9.30)

where C is a constant. Now, differentiating eq. (9.26) with respect to time, we have

$$\dot{\ell} = \dot{a}r,$$

(9.31)

and noting that the mass in the sub-volume can be expressed in terms of the density ρ as

$$M = \frac{4\pi}{3}\ell^3\rho = \frac{4\pi}{3}a^3r^3\rho,$$

(9.32)

we have

$$(\dot{a}r)^2 = \frac{2G}{ar}\frac{4\pi}{3}\rho(ar)^3 - C,$$

(9.33)

which we can rearrange by dividing through by $(ar)^2$ to obtain

$$\frac{\dot{a}^2}{a^2} = \frac{8\pi G\rho}{3} - \frac{C}{a^2r^2},$$

(9.34)

which is a special case of the first equation of motion that we seek. A correct derivation would yield

$$\frac{\dot{a}^2}{a^2} = \frac{8\pi G\rho}{3} - \frac{kc^2}{a^2R^2} + \frac{\Lambda}{3},$$

(9.35)

which is the first of the Friedmann–Lemaître equations. The principal oversimplification we made in the Newtonian derivation was in the definition of ρ, which we took to be the mass density. More generally, ρ is the mass-equivalent energy density of the Universe, including not only matter but radiation as well. The constants k and R are those of the Robertson–Walker metric; k is dimensionless, R is a physical length, and thus c^2/R^2 is in units of time^{-2}. A correct derivation would have also revealed that a second constant of integration is allowed, and this is the final term $\Lambda/3$. The constant Λ is known as the 'cosmological constant', and corresponds to the vacuum energy density of the Universe. It effectively acts as a repulsive term that counteracts the mutual gravitational attraction of the matter. We will take $\Lambda = 0$ throughout the rest of our discussion, given the absence of a reason to assume anything else. Equation (9.35) gives us the relationship between the expansion rate, energy density, and curvature of the Universe.

A second equation can be obtained from the first law of thermodynamics†

$$dE + PdV = TdS = 0 \tag{9.36}$$

where E is the *total* energy in a given comoving volume V (mass plus radiation) and PdV is the work done expanding the volume from V to $V + dV$. This equation states that the Universe is a *closed thermodynamic system*, i.e., that the total entropy, or given the homogeneity assumption the entropy per unit comoving volume, is constant. The radiation energy density is

$$\rho_{\mathrm{rad}}c^2 = \tilde{a}T^4, \tag{9.37}$$

where $\tilde{a} = 4\sigma/c = 7.56 \times 10^{-15}$ ergs cm^{-3} K^{-4} is the radiation density constant. The quantity ρ_{rad} is the mass equivalent of the radiation energy density.

Again considering small r, the total energy in a volume $V = (ar)^3$ is

$$E = \rho c^2 (ar)^3. \tag{9.38}$$

Since we are interested in the time rate of change of the total energy and pressure, we divide eq. (9.36) by dt to obtain

$$\frac{dE}{dt} + P\frac{dV}{dt} = 0,$$

$$\frac{d}{dt}\left(\rho c^2 a^3 r^3\right) + P\frac{d}{dt}\left(a^3 r^3\right) = 0,$$

$$a^3 r^3 c^2 \dot{\rho} + 3a^2 \dot{a} r^3 \rho c^2 + 3a^2 r^3 P \dot{a} = 0. \tag{9.39}$$

Solving this for $\dot{\rho}$ yields

$$\dot{\rho} = -\frac{3\dot{a}}{a}\left(\rho + \frac{P}{c^2}\right), \tag{9.40}$$

which is the second of the Friedmann–Lemaître equations.

We can relate the deceleration parameter to the physical variables through the Friedmann–Lemaître equations. We multiply eq. (9.35) by a^2 and differentiate with respect to time to obtain

$$2\dot{a}\ddot{a} = \frac{8\pi G}{3}\left(2\rho a \dot{a} + a^2 \dot{\rho}\right). \tag{9.41}$$

Using eq. (9.40) for $\dot{\rho}$ gives

$$2\dot{a}\ddot{a} = \frac{8\pi G}{3}\left[2\rho a \dot{a} - 3\dot{a}a\left(\rho + \frac{P}{c^2}\right)\right]. \tag{9.42}$$

---

† Equation (9.36) actually follows from the Einstein field equations and the perfect-fluid assumption (e.g., Weinberg 1972).

We now divide through by $2a\dot{a}$ to obtain

$$
\begin{aligned}
\frac{\ddot{a}}{a} &= \frac{4\pi G}{3}\left(2\rho - 3\rho - \frac{3P}{c^2}\right) \\
&= \frac{-4\pi G}{3}\left(\rho + \frac{3P}{c^2}\right).
\end{aligned} \tag{9.43}
$$

Recalling the definition of the deceleration parameter ($\ddot{a}/a = -qH^2$), we have

$$
qH^2 = \frac{4\pi G}{3}\left(\rho + \frac{3P}{c^2}\right). \tag{9.44}
$$

Let us now consider the case where $P/c^2 \ll \rho$, i.e., a 'pressureless' or 'matter-dominated' Universe. Then eq. (9.44) becomes

$$
qH^2 = \frac{4\pi G\rho}{3}. \tag{9.45}
$$

From eq. (9.35), we have

$$
\frac{8\pi G\rho}{3} = H^2 + \frac{kc^2}{R^2 a^2}. \tag{9.46}
$$

We can then combine these to obtain

$$
2qH^2 = H^2 + \frac{kc^2}{R^2 a^2} \tag{9.47}
$$

and thus

$$
2q = 1 + \frac{kc^2}{R^2 a^2 H^2}. \tag{9.48}
$$

This relates the sense of the curvature of the Universe as given by k to the deceleration parameter q. Equation (9.48) shows that $k = 0$ corresponds to $q_0 = 1/2$. We see that if $k = +1$, $q_0 > 1/2$, and $q_0 < 1/2$ for $k = -1$. Determination of the deceleration parameter is thus of fundamental importance since its magnitude relative to the critical value of $1/2$ (for $\Lambda = 0$) determines the geometry of space.

Evaluating these parameters at the present epoch, we can write

$$
\frac{kc^2}{R^2 a_0^2} = H_0^2(2q_0 - 1) \tag{9.49}
$$

and substituting into eq. (9.46) yields

$$
\frac{8\pi G\rho_0}{3} = 2q_0 H_0^2. \tag{9.50}
$$

The present value of the deceleration parameter q_0 depends on the present mass density ρ_0. The critical value $q_0 = 1/2$ represents a demarcation between an open and a closed Universe. The corresponding 'critical density', obtained by setting $k = 0$ and evaluating eq. (9.46) at the current epoch, is

$$\rho_{\text{crit}} \equiv \frac{3H_0^2}{8\pi G} \quad = \quad 1.9 \times 10^{-29} h_0^2 \text{ g cm}^{-3}$$
$$= \quad 2.8 \times 10^{11} h_0^2 \, M_\odot \text{ Mpc}^{-3}. \tag{9.51}$$

From eq. (9.45), it is clear that $q_0 < 1/2$ corresponds to $\rho_0 < \rho_{\text{crit}}$. It is convenient to define a dimensionless density parameter in terms of this critical density†

$$\Omega = \frac{\rho}{\rho_{\text{crit}}} = \frac{\rho_0}{3H_0^2/8\pi G} = 2q_0, \tag{9.52}$$

so that the present mass density can be written as

$$\rho_0 = \Omega \rho_{\text{crit}} = 1.9 \times 10^{-29} \, \Omega h_0^2 \text{ g cm}^{-3}. \tag{9.53}$$

Determination of the cosmological constants H_0, q_0, and Λ is obviously of fundamental importance to astrophysics, and involves levels of effort commensurate with their importance. Discussion of the current state of research in this area is beyond the scope of this book. It will suffice to say that the Hubble constant is almost certainly known to within a factor of around two, and lies somewhere in the range 50–100 km s^{-1} Mpc^{-1} (i.e., $0.5 \leq h_0 \leq 1.0$), with most recent investigations tending towards a value $h_0 \approx 0.7$. The density parameter Ω is more poorly known. A lower limit is set by the observed amount of luminous baryonic matter (i.e., galaxies) in the Universe. Assuming nominal mass-to-light ratios, we find that $\Omega \approx 0.1$, with considerable margin for error. On the other hand, there are theoretical and philosophical reasons for preferring a value $\Omega = 1$, although this leaves the whereabouts and nature of 90% of the mass of the Universe unaccounted for (one manifestation of the well-known 'dark-matter' problem). In eq. (9.53), our ignorance is parameterized in the product Ωh_0^2, which may have values anywhere in the range $0.025 \lesssim \Omega h_0^2 \lesssim 1$. When specific estimates are desirable, we will choose the case $q_0 = 1/2$; in the absence of more definitive evidence, the simplest choice is the $\Omega = 1$, $\Lambda = 0$ solution, which is known as the 'Einstein–de Sitter' model.

9.3 Time Dependence of Cosmological Parameters

In this section, we consider the relationships between the cosmological parameters q and H and the expansion parameter a, redshift z, and cosmic time t.

We will first consider how the energy density varies with a. Multiplying eq. (9.40) by a^3, we have

$$a^3 \frac{d\rho}{dt} + 3a^2 \dot{a} \left(\rho + \frac{P}{c^2} \right) = 0. \tag{9.54}$$

† More generally, $\Omega = 2q_0 + \Lambda/3H_0^2$.

Note that

$$\frac{d}{dt}\left(a^3\rho\right) = a^3\dot{p} + 3a^2\dot{a}\rho, \tag{9.55}$$

which is eq. (9.54) for the case of a matter-dominated Universe, i.e., $P \ll \rho c^2$. This shows that the quantity $a^3\rho$ must be constant with time, so $\rho \propto a^{-3} \propto (1+z)^3$. This is in fact what we expect naïvely from the Hubble law – as the distance scale decreases in the past like $1 + z$, comoving volumes must scale as $(1+z)^{-3}$. If there are not sources or sinks for matter in the Universe, the mass per comoving volume is constant. The mass per *proper* volume is thus higher in the past by a factor $(1+z)^3$.

We can also consider the case of a 'radiation-dominated' Universe, in which $P = \rho c^2/3$, a familiar result from thermodynamics. In this case, eq. (9.54) becomes

$$a^3\frac{d\rho}{dt} + 4a^2\dot{a}\rho = 0. \tag{9.56}$$

Multiplying through by a, we have

$$a^4\dot{p} + 4a^3\dot{a}\rho = 0 \tag{9.57}$$

which is equivalent to

$$\frac{d}{dt}\left(a^4\rho\right) = 0. \tag{9.58}$$

Thus, in a radiation-dominated Universe, $a^4\rho$ is constant, so $\rho \propto a^{-4} \propto (1+z)^4$. The explanation for this is quite simple. As in the matter-dominated case, the number of photons per unit comoving volume is constant, so their proper density increases with redshift as $(1+z)^3$. However, on account of the cosmological redshift (eq. 9.14) the energy of each photon also scales as $(1+z)$, so the *energy* density in photons (i.e., number density times energy per photon) is proportional to $(1+z)^4$.

We also wish to investigate how the cosmological scale constants, H_0 and q_0, vary with the expansion parameter a. From eqs. (9.35) and (9.46), we have

$$\frac{kc^2}{R^2a^2} = H^2(2q-1) = \frac{8\pi G\rho}{3} - \frac{\dot{a}^2}{a^2}. \tag{9.59}$$

Normalizing this by dividing through by the present value of this equation gives

$$\frac{a_0^2}{a^2} = \left(\frac{8\pi G\rho}{3} - \frac{\dot{a}^2}{a^2}\right) \frac{1}{H_0^2(2q_0 - 1)} \tag{9.60}$$

By using eqs. (9.14) and (9.50) and noting that we can write the energy density at a previous epoch during the matter-dominated era as $\rho = \rho_0(1+z)^3$, this becomes

$$H_0^2(2q_0 - 1)(1+z)^2 = 2q_0 H_0^2(1+z)^3 - \frac{\dot{a}^2}{a^2}, \tag{9.61}$$

which can be rearranged to

$$\frac{\dot{a}^2}{a^2} = H_0^2(1+z)^2(1+2q_0 z) \tag{9.62}$$

The quantity $\dot{a}/a$ can be readily identified as the Hubble parameter at the epoch corresponding to redshift z, so we can write this as

$$H(z) = H_0 (1 + z)(1 + 2q_0 z)^{1/2}. \tag{9.63}$$

We can also compute how q_0 varies with redshift z. Using eq. (9.50) and $\rho = \rho_0(1+z)^3$, we have

$$\frac{8\pi G\rho}{3} = 2q_0 H_0^2 (1 + z)^3, \tag{9.64}$$

and thus eq. (9.59) becomes

$$2qH^2 = 2q_0 H_0^2 (1 + z)^3. \tag{9.65}$$

Combining this with eq. (9.63), we have

$$q(z) = \frac{q_0 (1 + z)}{1 + 2q_0 z}. \tag{9.66}$$

We now have simple mathematical relationships between the observable parameters H_0, q_0, and z and the expansion parameter a. The remaining task is to relate a and cosmic time t directly. First, note that the second term in eq. (9.35) can be written

$$\frac{kc^2}{R^2 a^2} = \frac{kc^2}{R^2 a_0^2} \frac{a_0^2}{a^2} = \frac{a_0^2}{a^2} H_0^2 (2q_0 - 1) \tag{9.67}$$

and

$$\frac{8\pi G\rho}{3} = \frac{8\pi G\rho_0}{3} \frac{a_0^3}{a^3} = \frac{a_0^3}{a^3} 2q_0 H_0^2. \tag{9.68}$$

Thus, eq. (9.35) becomes

$$\frac{\dot{a}^2}{a^2} = \frac{a_0^3}{a^3} 2q_0 H_0^2 - \frac{a_0^2}{a^2} H_0^2 (2q_0 - 1). \tag{9.69}$$

Multiplying this through by a^2/a_0^2, we have

$$\frac{\dot{a}^2}{a_0^2} = \frac{a_0}{a} 2q_0 H_0^2 - H_0^2 (2q_0 - 1) = H_0^2 \left[1 - 2q_0 + \frac{2q_0 a_0}{a} \right]. \tag{9.70}$$

Taking the square root of this equation and reorganizing, we get

$$dt = \frac{da}{H_0 a_0 \left[1 - 2q_0 + 2q_0 a_0/a \right]^{1/2}}. \tag{9.71}$$

We can now compute $a(t)$ by integrating this equation, i.e.,

$$\int_0^t dt = t = \frac{1}{a_0 H_0} \int_0^a \frac{da}{\left[1 - 2q_0 + 2q_0 a_0/a \right]^{1/2}}. \tag{9.72}$$

To simplify this, let $x = a_0/a$ and $dx = -a_0 da/a^2$, so

$$t = \frac{1}{H_0} \int_{a_0/a}^{\infty} \frac{dx}{x^2 \left[1 - 2q_0 + 2q_0 x \right]^{1/2}}, \tag{9.73}$$

which gives us the age of the Universe (i.e., cosmic time t) as a function of q_0 and H_0. The trivial case is the Einstein–de Sitter solution, where

$$t = \frac{1}{H_0} \int_{a_0/a}^{\infty} \frac{dx}{x^{5/2}} = \frac{-2}{3H_0} x^{-3/2} \Big|_{a_0/a}^{\infty} = \frac{2}{3H_0} \left(\frac{a}{a_0}\right)^{3/2} \tag{9.74}$$

or

$$\frac{a}{a_0} = \left(\frac{3H_0 t}{2}\right)^{2/3}. \tag{9.75}$$

The present age of the Universe is given by $a = a_0$, so $t = 2/3H_0$ ($< 1/H_0$) for $q_0 = 1/2$. Note that we could have arrived at this special result from eq. (9.35) by setting $k = \Lambda = 0$ and using $\rho = \rho_0 (a_0/a)^3$.

For arbitrary values of q_0, the integral is somewhat more complicated. It is straightforward to show that for $q_0 > 1/2$, the solution is

$$t = \frac{q_0}{H_0(2q_0 - 1)^{3/2}} (\theta - \sin \theta), \tag{9.76}$$

where

$$\theta = \cos^{-1} \left[1 - \left(\frac{2q_0 - 1}{q_0}\right) \frac{a}{a_0}\right]. \tag{9.77}$$

For the case $q_0 < 1/2$, the solution is

$$t = \frac{q_0}{H_0(1 - 2q_0)^{3/2}} (\sinh \psi - \psi), \tag{9.78}$$

where

$$\psi = \cosh^{-1} \left[1 + \left(\frac{1 - 2q_0}{q_0}\right) \frac{a}{a_0}\right]. \tag{9.79}$$

Again, the present age of the Universe is given by setting $a = a_0$.

An interesting and especially relevant quantity that we can now compute is the age of an object at arbitrary redshift as we currently observe it. The age of the Universe at the present time is given by integrating from 0 to t_0. Similarly, the age of the Universe at the time a distant object is emitting photons in our direction is given by integrating from 0 to t_1. Thus light-travel time for the photons, or 'lookback time' τ, is the difference between these, i.e., the integral in eq. (9.73) from t_1 to t_0,

$$\begin{aligned} \tau = t_0 - t_1 &= \frac{1}{a_0 H_0} \int_{a_1}^{a_0} \frac{da}{[1 - 2q_0 + 2q_0 a_0/a]^{1/2}} \\ &= \frac{1}{H_0} \int_{1}^{a_0/a_1} \frac{dx}{[1 - 2q_0 + 2q_0 x]^{1/2} x^2} \end{aligned} \tag{9.80}$$

For $q_0 = 1/2$,

$$\tau = \frac{1}{H_0} \int_{1}^{a_0/a_1} \frac{dx}{x^{5/2}} = \frac{2}{3H_0} \left[1 - \frac{1}{(1 + z)^{3/2}}\right]. \tag{9.81}$$

For $q_0 > 1/2$,

$$\tau = \frac{q_0}{H_0 \, (2q_0 - 1)^{3/2}} \, [\theta - \sin\theta] \Big|_{\theta(z)}^{\theta(z=0)}, \tag{9.82}$$

where

$$\theta(z) = \cos^{-1}\left[1 - \frac{2q_0 - 1}{q_0\,(1+z)}\right]. \tag{9.83}$$

For $q_0 < 1/2$,

$$\tau = \frac{q_0}{H_0 \, (1 - 2q_0)^{3/2}} \, [\sinh\psi - \psi] \Big|_{\psi(z)}^{\psi(z=0)}, \tag{9.84}$$

where

$$\psi(z) = \cosh^{-1}\left[1 + \frac{1 - 2q_0}{q_0\,(1+z)}\right]. \tag{9.85}$$

We give these somewhat complicated formulae explicitly as they are of considerable use in observational cosmology. Lookback time τ is shown as a function of z and q_0 in Fig. 9.2.

We can complete this set of relationships by computing the distance parameters r_1 and $f(r_1)$ as function of the observables z, q_0, and H_0. From the definition of $f(r_1)$, eq. (9.7),

$$f(r_1) = c \int_{t_1}^{t_0} \frac{dt}{a(t)} = \int_0^r \frac{dr}{\left(1 - kr^2/R^2\right)^{1/2}}. \tag{9.86}$$

We can write

$$c \int_{t_1}^{t_0} \frac{dt}{a(t)} = c \int_{a_1}^{a_0} \frac{da}{a\dot{a}}, \tag{9.87}$$

and combine this with eq. (9.71), which we rearrange as

$$\dot{a} = a_0 H_0 \left[1 - 2q_0 + 2q_0 a_0/a\right]^{1/2} \tag{9.88}$$

to obtain

$$f(r_1) = \frac{c}{a_0 H_0} \int_{1/(1+z)}^1 \frac{dx}{[1 - 2q_0 + 2q_0 x]^{1/2} \, x}. \tag{9.89}$$

Regardless of the value of k, the solution of this integral yields the result

$$r_1 = \frac{c}{a_0 H_0 q_0^2} \left[\frac{z q_0 + (q_0 - 1)(-1 + \sqrt{2q_0 z + 1})}{1 + z}\right]. \tag{9.90}$$

With this formula, it is now possible to determine the magnitude of various distance measures in terms of the observables, and to translate angular sizes to linear dimensions. Consider, for example, proper distances $d_p = a_0 f(r_1)$ for the simple case of $q_0 = 1/2$.

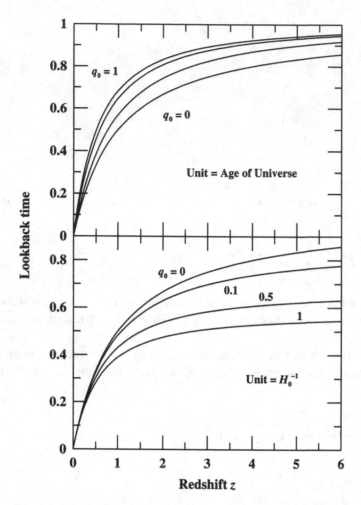

Fig. 9.2. The 'lookback time' as a function of redshift z for various values of q_0. The lookback time is the travel time to the observer for photons originating in an object of observed redshift z. In the upper panel, the lookback time is measured as a fraction of the current age of the Universe. In the lower panel, it is given in units of the Hubble time H_0^{-1} for easy translation into units of time. Adapted from Peebles (1971).

In this case,

$$d_p = a_0 r_1 = \frac{2c}{H_0}\left[1 - \frac{1}{(1+z)^{1/2}}\right]. \tag{9.91}$$

We note that if z is small, $(1+z)^{-1/2} \approx 1 - z/2$, so in this limit the proper distance becomes

$$d_p \approx \frac{cz}{H_0}, \tag{9.92}$$

as expected.

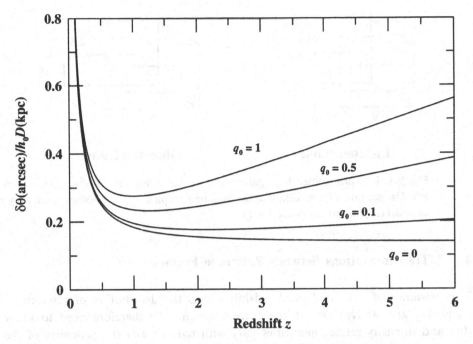

Fig. 9.3. The relationship between the angular ($\delta\theta$) and linear (D) sizes of objects at redshift z for various values of q_0, as in eq. (9.94). Adapted from Peebles (1971).

It is an interesting exercise to compute the angular diameter of an object of linear diameter D (say, a standard galaxy) as a function of z and q_0. From eq. (9.24),

$$\delta\theta = \frac{D}{d_A} = \frac{D}{ar_1} = \frac{H_0 q_0^2}{c} \frac{(1+z)^2}{\left[zq_0 + (q_0-1)(-1+\sqrt{2q_0z+1})\right]} D. \tag{9.93}$$

In units appropriate for distant galaxies, this can be written

$$\delta\theta('') = 0.0688h_0 \frac{q_0^2(1+z)^2}{\left[zq_0 + (q_0-1)(-1+\sqrt{2q_0z+1})\right]} D(\text{kpc}). \tag{9.94}$$

This relationship is shown in Fig. 9.3. One of the interesting aspects of this figure is that for sufficiently large values of q_0, there is some value of z for which the angular size is a minimum, and for larger z, the angular sizes increase. The explanation of this non-intuitive result is that the existence of even a small amount of mass in the Universe has significantly decelerated the cosmological expansion over the Hubble time $1/H_0$; in the Universe the expansion rate was so much larger that, for some time in the early history of the Universe, the proper distances between us and distant objects (at fixed comoving coordinates) were increasing faster than the speed of light, which means that for a time, photons emitted in our direction were actually *receding* from us.

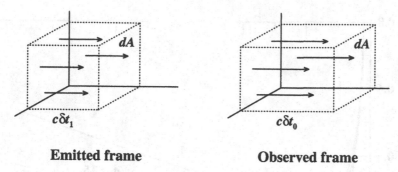

Emitted frame   **Observed frame**

Fig. 9.4. Photons emitted by a galaxy at t_1 occupy a volume $dA\,cdt_1$, as shown on the left. On account of time dilation, the volume occupied by the photons when they reach us is $dA\,cdt_0$, where $dt_0 = dt_1(1 + z)$.

9.4   Transformations Between Reference Frames

The existence of a cosmological redshift affects the distribution of photons in both frequency and arrival rate at the current epoch. We therefore need to know how flux and intensity-related quantities vary with redshift and the geometry of the Universe.

9.4.1   Redshifted Spectra

Consider as a simple case a swarm of photons emitted in our direction by a galaxy at some cosmic time t_1, and these photons are arriving here at the present time t_0. Initially, suppose that these photons occupy a volume $dA\,cdt_1$, as shown in Fig. 9.4.

In the rest frame of the observer, the rate at which these photons arrive is decreased by time dilation by a factor $1 + z$. The photons, when they arrive at the observer, are spread out over the larger volume $dA\,cdt_0$, where $dt_0 = dt_1(1 + z)$. An equivalent way to look at this is that the length of the box along the line of sight is Lorentz-contracted in the observed frame by a factor of $1 + z$. As shown earlier, the frequencies of each of the photons arriving at the observer are decreased by a factor $1 + z$, which is of course what defines the redshift.

The number density of photons in the box in the frequency range v to $v + dv$ which are moving in the $-r$ direction and cross the surface dA per second in Fig. 9.4 is given by n_v photons s^{-1} cm^{-2} Hz^{-1}. The total number of photons at frequency v_1 (i.e., in the range v_1 to $v_1 + dv_1$) at t_1 is thus

$$n_v(v_1)\,dA\,dv_1\,cdt_1. \tag{9.95}$$

The number of photons in the box must be conserved despite the fact that the box changes size by the time of arrival at t_0. To see this, we note that the total number of photons at observed frequency v_0 (i.e., in the range v_0 to $v_0 + dv_0$ is

$$
\begin{aligned}
n_v(v_0)\, dA\, dv_0\, cdt_0 &= n_v(v_0) \frac{dv_1}{1+z}\, dA\, cdt_1(1+z) \\
&= n_v(v_0)\, dA\, dv_1\, cdt_1,
\end{aligned}
\tag{9.96}
$$

so from eq. (9.95), we see that

$$
n_v(v_0) = n_v(v_1).
\tag{9.97}
$$

Thus, the *number* of photons is preserved, but the photons are shifted in frequency from the range v_1 to $v_1 + dv_1$ to v_0 to $v_0 + dv_0$ and their arrival rate has been decreased.

Of more interest is the *energy flux* (ergs s^{-1} cm^{-2} Hz^{-1}) due to photons crossing surface dA in the $-r$ direction. Since n_v is proportional to the *photon flux* through the surface dA, the energy flux is easily obtained by multiplying n_v by the energy per photon hv. Using the prime symbol (') to denote properties measured in the *observer's* frame, we have

$$
\begin{aligned}
F'_v(v_0) &= hv_0 n_v(v_0) \\
&= h\frac{v_1}{1+z} n_v(v_1) \\
&= \frac{F_v(v_1)}{1+z}.
\end{aligned}
\tag{9.98}
$$

If the photons are distributed over some range of frequencies, the *total* energy flux is given by integrating over frequency, so

$$
\begin{aligned}
F' &= \int F'_v(v_0) dv_0 \\
&= \int \frac{F_v(v_1)}{1+z} \frac{dv_1}{1+z} \\
&= \frac{F}{(1+z)^2}.
\end{aligned}
\tag{9.99}
$$

This is perhaps the most intuitive of these transformations, since if we consider a monochromatic flux of photons, we find that their *energy flux* in the observer's frame is decreased by a factor of $(1+z)^2$, where one factor of $1+z$ is due to the decreased energy per photon and the other factor of $1+z$ is due to their decreased arrival rate as the source recedes from the observer.

It is often preferable to work in wavelength-dependent (F_λ) units rather than frequency-dependent (F_v) units. The transformation between them is trivial (eq. 1.6) since no matter how one measures the flux, the energy must be the same. Since

$F_\lambda = F_\nu c / \lambda^2$, the transformation between observed and rest frames becomes

$$
\begin{aligned}
F_\lambda'(\lambda_0) &= \frac{F_\nu'(\nu_0) c}{\lambda_0^2} \\
&= \frac{F_\nu(\nu_1)}{1+z} \frac{c}{\lambda_1^2 (1+z)^2} \\
&= \frac{F_\lambda(\lambda_1)}{(1+z)^3}.
\end{aligned}
\tag{9.100}
$$

Finally, line equivalent widths (eq. 1.10) transform as wavelengths, i.e.,

$$
W_{\lambda_0}' = \frac{F_{\text{line}}'}{F_c'(\lambda_0)} = \frac{F_{\text{line}}}{(1+z)^2} \frac{(1+z)^3}{F_c(\lambda_1)} = W_{\lambda_1}(1+z).
\tag{9.101}
$$

It is important to note that these transformations do not explicitly include effects of a purely cosmological origin; they apply not only to spectra redshifted by cosmological expansion, but to ordinary Doppler-shifted spectra as well.

9.4.2   Geometrical Effects

We now suppose that a distant source at comoving coordinate r_1 has some luminosity L. From the Robertson–Walker metric, we see that when this light has reached the origin, it is spread over a sphere of surface area $4\pi (a_0 r_1)^2$, so the flux one would measure in the same reference frame is

$$
F_1 = \frac{L}{4\pi (a_0 r_1)^2}.
\tag{9.102}
$$

The flux actually measured by the observer at the origin, corrected for the Doppler effect as per the last section, is

$$
F_0 = \frac{F_1}{(1+z)^2} = \frac{L}{4\pi (1+z)^2 (a_0 r_1)^2} = \frac{L}{4\pi d_L^2},
\tag{9.103}
$$

where, in analogy to the Euclidean formula for geometrical dilution, we have defined a 'luminosity distance' $d_L = (1+z) a_0 r_1$.[†] From the above equations and eq. (9.90),

$$
d_L = \frac{c}{H_0 q_0^2} \left[z q_0 + (q_0 - 1) \left(-1 + \sqrt{2 q_0 z + 1} \right) \right].
\tag{9.104}
$$

It is left as an exercise to show that for small values of z

$$
\begin{aligned}
d_L &= \frac{cz}{H_0} (1 + z/2) \qquad (q_0 = 0) \\
&\approx \frac{cz}{H_0} (1 + z/4) \qquad (q_0 = 1/2).
\end{aligned}
\tag{9.105}
$$

Unfortunately, eq. (9.104) is not always easy to use because it tends to be numerically unstable at small values of z and q_0. We therefore recommend using the equivalent

---

† The quantity d_L might more properly be called a '*bolometric* luminosity distance', since it refers specifically to the flux (and luminosity) integrated over some wavelength or frequency range.

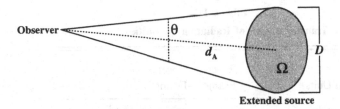

Fig. 9.5. The extended source geometry assumed in the derivation of the surface-brightness transformation. A source of linear diameter D subtends an angle θ and solid angle Ω as seen by an observer at distance d_A.

form (Terrell 1977)

$$d_L = \frac{cz}{H_0}\left[1 + \frac{z(1-q_0)}{(\sqrt{2q_0z+1}+1+zq_0)}\right].\tag{9.106}$$

We can also consider the surface brightness of an extended source (as in Fig. 9.5) of angular size $\delta\theta$, linear size D, and luminosity L. The solid angle subtended by the source is $\delta\Omega = \pi\,\delta\theta^2/4$, and the relationship between angular size $\delta\theta$ and linear size D is given by eq. (9.24). Thus

$$\delta\Omega = \frac{\pi D^2}{4d_A^2} = \frac{\pi D^2(1+z)^2}{4a_0^2r_1^2}.\tag{9.107}$$

The surface brightness of the source (ergs s^{-1} cm^{-2} arcsec^{-2}) is thus

$$
\begin{aligned}
\Sigma' &= \frac{F'}{\delta\Omega} = \frac{L}{4\pi a_0^2 r_1^2(1+z)^2}\frac{4a_0 r_1^2}{\pi D^2(1+z)^2}\\
&= \frac{L}{\pi^2 D^2(1+z)^4}\\
&= \frac{\Sigma}{(1+z)^4}.
\end{aligned}\tag{9.108}
$$

It is obvious from the units of surface brightness that *intensity* scales in the same way, as $(1+z)^{-4}$. Since the total intensity is the integral of the specific intensity, $I = \int I_\nu d\nu$, and $d\nu_0 = d\nu_1/(1+z)$, the specific intensity must transform like

$$I'_\nu(\nu_0) = \frac{I_\nu(\nu_1)}{(1+z)^3} = \frac{I_\nu(\nu_0[1+z])}{(1+z)^3}.\tag{9.109}$$

An important result of this transformation can be seen by considering the observed intensity of a blackbody spectrum characterized by temperature T_1 in the rest frame, and redshifted by z. In the rest frame, the specific intensity is

$$I_\nu(\nu_1) = \frac{2h\nu_1^3}{c^2}\left[\exp\left(\frac{h\nu_1}{kT_1}\right)-1\right]^{-1}.\tag{9.110}$$

Table 9.1

Transformation of Radiation Quantities

Quantity (Common Units)	Rest Frame	Observed Frame
Frequency (Hz)	ν_1	$\nu_0 = \nu_1(1+z)^{-1}$
Wavelength (Å)	λ_1	$\lambda_0 = \lambda_1(1+z)$
Specific Flux (ergs s^{-1} cm^{-2} Hz^{-1})	$F_\nu(\nu_1)$	$F'_\nu(\nu_0) = F_\nu(\nu_1)/(1+z)$
Specific Flux (ergs s^{-1} cm^{-2} Å^{-1})	$F_\lambda(\lambda_1)$	$F'_\lambda(\lambda_0) = F_\lambda(\lambda_1)/(1+z)^3$
Flux (ergs s^{-1} cm^{-2})	F	$F' = F/(1+z)^2$
Surface Brightness (ergs s^{-1} cm^{-2} arcsec^{-2})	Σ	$\Sigma' = \Sigma/(1+z)^4$
Specific Intensity (ergs s^{-1} cm^{-2} Hz^{-1} ster^{-1})	$I_\nu(\nu_1)$	$I'_\nu(\nu_0) = I_\nu(\nu_1)/(1+z)^3$
Intensity (ergs s^{-1} cm^{-2} ster^{-1})	I	$I' = I/(1+z)^4$
Equivalent Width (Å)	$W(\lambda_1)$	$W'(\lambda_0) = W(\lambda_1)(1+z)$

In the observed frame, the specific intensity is thus

$$
\begin{aligned}
I'_\nu(\nu_0) &= \frac{I_\nu[\nu_0(1+z)]}{(1+z)^3} \\
&= \frac{1}{(1+z)^3} \frac{2h\nu_0^3(1+z)^3}{c^2}\left[\exp\left(\frac{h\nu_0(1+z)}{kT_1}\right) - 1\right]^{-1} \\
&= \frac{2h\nu_0^3}{c^2}\left[\exp\left(\frac{h\nu_0}{kT_0}\right) - 1\right]^{-1},
\end{aligned}
\tag{9.111}
$$

where we have defined $T_0 = T_1/(1+z)$. We thus see that a redshifted blackbody is still a blackbody, but one characterized by a temperature decreased by the factor $1+z$. This is the reason that the cosmic background radiation retains a blackbody form.

The transformations of various quantities between the observed frame and rest frame are summarized for future use in Table 9.1.

10 Quasar Surveys

One of the main goals of QSO research is to use these objects as a probe of the history of the Universe. Two specific aims are first, to determine the characteristics of the QSO population as a function of redshift, and second, to find the lookback time at which QSOs first appeared, as this provides some measure of the time scale for galaxy formation in the early Universe. Both of these important aims require large and preferably unbiased samples of QSOs. In this chapter, we consider how large samples of QSOs might be obtained through various survey techniques, and how the samples we obtain might be affected by various biases.

The measurable quantity that will result from surveys is the QSO 'surface density' $d\mathcal{N}(F,z)/d\Omega$, i.e., number of QSOs per unit solid angle (square degree) as a function of flux F and redshift z. From this, we can compute the 'luminosity function', which is the relative number of AGNs at a given luminosity, and the 'space density', which is the total number of sources per unit comoving volume† over some specified luminosity range – when the luminosity function is correctly normalized, the total space density is simply the integral of the luminosity function over its entire range.

The primary goal of QSO surveys then is to determine $d\mathcal{N}(F,z)/d\Omega$ in an accurate and unbiased fashion. This is a difficult and complicated undertaking because QSOs are faint and their surface density is low; the total surface density of QSOs brighter than $B = 21$ mag is only $\sim 40\,\mathrm{deg}^{-2}$. Even at the comparatively sparsely populated Galactic poles the integrated surface density of stars is about $1600\,\mathrm{deg}^{-2}$ to the same limiting magnitude. Obviously, candidate QSOs must be preselected on some basis prior to investing many hours of large telescope time obtaining spectra of faint sources. The optimum strategy is to preselect QSO candidates in some way that minimizes 'contamination' of the candidate list by effectively excluding stars and at the same time minimizes the probability that real QSOs will be inadvertently excluded. Clearly very large numbers of QSOs are required to determine how the population changes with z, and as we shall see, we must pay particular attention to selection effects which might give us a misleading picture.

We first recall briefly some of the most important ways that QSOs have been found:

Radio position: The first QSOs were discovered by the coincidence of radio sources with star-like optical sources. The principal drawbacks to this approach are (a) that

---

† Note, however, that comoving densities are always referred to equivalent *proper volumes* at the current epoch, i.e., densities measured at redshift z are decreased by a factor $(1 + z)^3$ and given in units such as number of sources per Gpc^3.

relatively high-accuracy (by the standards of single-dish telescopes) radio positions ($\sim 1'$) are required for an unambiguous identification, and (b) that this isolates only *radio-loud* QSOs which are not only relatively rare (about 5–10% of all AGNs), but may not be very representative of the AGN population as a whole.

Radio position + UV excess: An efficient way to find individual QSOs is to examine the fields of radio sources with relatively poor (several arcminutes) positional accuracy (for example, the Ohio sources) and obtain spectra of only the bluest objects in the field. These objects can be readily identified by a process as simple as comparing the red and blue Palomar Sky Survey photographs. Again, this method focuses on radio-loud objects and, as will become apparent below, is hopelessly entangled with redshift-dependent selection effects that render the resulting data nearly useless for determining anything about QSO evolution. This technique was very popular in the first decade of QSO research, when the number of known QSOs was relatively small and one of the principal interests was finding individual very high-redshift objects.

Colors: Traditionally in QSO research, this has meant 'UV-excess' objects, as mentioned above. Developments in detector technology and data processing capabilities have led to similar strategies based on multicolor data which more effectively isolate QSOs from starlight-dominated objects.

Slitless spectroscopy: The idea behind slitless spectroscopy is to increase the survey efficiency by obtaining spectra of a large number of objects in a single field at once, for example, by use of an objective prism in front of a wide-field telescope. QSO SEDs and strong emission lines make their spectra distinct from those of stars, even at very low spectral resolution. The principal problem with this method is that the signal-to-noise ratio of slitless spectra is inherently limited.

X-Ray emission: Strong X-ray emission is such a common property of AGNs that it is probably the best way to obtain the most complete census of AGNs, at least at low redshift – the largest space density of Seyfert 1 galaxies is found through observations in the 2–10 keV range (Mushotzky, Done, and Pounds 1993). The principal limitation is that both limited flux sensitivity and angular resolution make it difficult to identify all but relatively nearby or extremely luminous AGNs.

This list is not comprehensive. QSOs have been identified on the basis of other properties, such as optical variability or high infrared luminosity (e.g., Low *et al.* 1988), and even serendipity (e.g., McCarthy *et al.* 1988). But in order to minimize biases in selection and maximize the number of sources found, we will focus on the most general and efficient methods: color selection and slitless spectroscopy. It is important to realize that color selection and some slitless methods yield only lists of *candidate* QSOs. Once the QSO candidates in a field have been isolated, follow-up spectroscopy is required (a) to verify that the candidates are indeed QSOs, and (b) to measure their redshifts accurately. Modern methods for isolating candidates yield success rates (i.e., the percentage of candidate objects that turn out to be QSOs with measurable redshifts) which are reasonably high (~40–80%). The high efficiencies of these methods are directly attributable to the fact that modern QSO survey data are in a digital format, in contrast to the early surveys that relied primarily on visual inspection of photographic materials. Quantitative selection criteria can be applied in a straightforward and completely reproducible fashion by computer codes, and the selection algorithms can be extensively tested through Monte Carlo techniques to determine their efficacy. One clear vestige of the days of photographic surveys is that most optical results are referred to the *B* band, which is the Johnson photometric band that is closest to the response of the sensitive Kodak IIIaJ emulsion that has been used in much of the photographic survey work.†

10.1 Basic Principles of QSO Surveys

The efficiency with which QSOs can be detected by a particular observing strategy is a function of several parameters, the most important and easily identifiable being luminosity, redshift, and SED (see Hewett and Foltz 1994 for a comprehensive discussion), and these are the independent variables that can be used to define a 'survey selection function'. Before we elaborate on this concept, it is worthwhile to consider how we expect the surface density of QSOs to depend on the minimum flux or limiting apparent magnitude that we can detect, since this is the most important factor in detection of QSOs.

10.1.1 Expected Number Counts

A common goal of astronomical surveys is to obtain a 'flux-limited sample' of sources, i.e., to detect all of the sources in a given region in the sky with fluxes higher than some detection limit *S*. For simplicity, we consider the case of a population of objects which all have the same luminosity *L*, and suppose that space is Euclidean. The space density of objects, as a function of distance *r* from the observer is $n(r)$. The total number of sources that will be found in a volume element dV is

$$d\mathcal{N}(r) = n(r)\,dV = n(r)r^2 dr\,d\Omega, \tag{10.1}$$

so the surface density of sources, i.e., the number of sources per unit solid angle

---

† Some of the photographic measurements of QSO magnitudes have been expressed in so-called *J* (or sometimes B_J) magnitudes, based on the response of the IIIaJ emulsion combined variously with either a Schott GG-385 or GG-395 filter; *J* magnitudes are very close to Johnson *B*, but with a slight redward extension. Care should be taken not to confuse this with the broad-band Johnson *J* filter, whose effective wavelength is ~1.25 μm. Similarly, the Kodak IIIaF emulsion (also known as 127-02) used with a GG-495 filter defines an *F* magnitude, which is approximately equivalent to $(V + R)/2$ on the Johnson system (see Koo, Kron, and Cudworth 1986).

between r and $r + dr$, is

$$\frac{d\mathcal{N}(r)}{d\Omega} = n(r)r^2dr. \tag{10.2}$$

The observed flux from a particular source at r is $F = L/4\pi r^2$. Now suppose that we detect all sources in a given field down to some limiting flux S, i.e., all sources with $F \geq S$ are counted. This means that we detect sources out to some maximum distance

$$r_{max} = \left(\frac{L}{4\pi S}\right)^{1/2}. \tag{10.3}$$

The total number of sources per unit solid angle detected above the limiting flux S is thus

$$N(S) = \int \frac{d\mathcal{N}}{d\Omega}(F \geq S) = \int \frac{d\mathcal{N}}{d\Omega}(r \leq r_{max}) = \int_0^{r_{max}} n(r)r^2dr. \tag{10.4}$$

The quantity $N(S)$ is thus the *cumulative* distribution of sources as a function of flux. For the case of a uniform density of objects $n(r) = n_0$, this becomes

$$N(S) = n_0\frac{r_{max}^3}{3} = \frac{n_0}{3}\left(\frac{L}{4\pi S}\right)^{3/2}, \tag{10.5}$$

and taking the logarithm of this gives

$$\log N(S) = \log\left[\frac{n_0 L^{3/2}}{3(4\pi)^{3/2}}\right] - \frac{3}{2}\log S. \tag{10.6}$$

We see that the number of sources brighter than some flux S should be proportional to $S^{-3/2}$ for a constant density in Euclidean space. Indeed the observational realization of this relationship is known as the 'log N – log S test' which in principle can be used to test the hypothesis that the observed population has a constant space density. In this form, the log N – log S test was developed to test for the evolution of radio sources. The beauty of the test is that distances to individual sources are not required.

In the optical version of this test, we consider sources brighter than a limiting apparent magnitude $m \propto -2.5\log S$, so $\log S \propto -0.4m$. The second term in eq. (10.6) is

$$-\frac{3}{2}\log S = (-1.5)(-0.4m) = 0.6m. \tag{10.7}$$

Thus, expressed on the magnitude system commonly used in optical astronomy, the number of sources brighter than m is

$$\log N(m) \propto 0.6m. \tag{10.8}$$

For a constant space density, the number of objects detected increases by a factor of $10^{0.6} \approx 4$ with each magnitude. This immediately leads us to the sobering realization that for a constant space density of QSOs, 80% of our sample will be within 1 mag of our survey detection limit.

If the sources have some distance-independent distribution in luminosity L, the first, constant term in eq. (10.6) changes, but the dependence on S is unchanged. We must make the assumption that the density of objects as a function of distance and luminosity can be separated into two parts,

$$n(r, L)\, dr\, dL = n(r)\, \phi(L)\, dr\, dL, \tag{10.9}$$

where $\phi(L)\, dL$ is the luminosity function. In this case, again assuming that $n(r) = n_0$, eq. (10.4) becomes

$$N(S) = \int\!\!\int_0^{r_{\max}(L)} n(r, L) r^2 \, dr \, dL = \frac{n_0}{3} (4\pi S)^{-3/2} \int L^{3/2} \phi(L)\, dL. \tag{10.10}$$

Thus, a straightforward test of whether or not the space density of a particular class of object is constant is to plot as a function of either the logarithm of limiting flux $\log S$ or the limiting apparent magnitude m, the logarithm of the cumulative distribution and measure the slope. A slope $d \log N(S)/d \log S = -3/2$ (or $d \log N(m)/dm = 0.6$) is expected if the space density is constant. A steeper dependence ($d \log N(S)/d \log S < -1.5$ or $d \log N(m)/dm > 0.6$), for example, would be taken as an indication that the space density increases with r. Generalization of the $\log N - \log S$ test to non-Euclidean geometries is straightforward, and will be discussed in Chapter 11.

10.1.2 Problems with Flux-Limited Samples

In practice, there are a number of serious difficulties with the $\log N - \log S$ test. First of all, there is the implicit assumption that the *shape* of the luminosity function is the same everywhere. This does not in fact seem to be true for QSOs; the luminosity function is considerably different now than it was at $z \gtrsim 2$. This is a particular problem with sources such as AGNs which span a very large range in luminosity, since then the objects which contribute to the total counts at a given flux are distributed over a very wide range in distance.

The most important problem with the $\log N - \log S$ test is that the results are very sensitive to 'completeness', i.e., at least in principle, *all* sources with fluxes higher than the limit S must be detected. If sources are missed, which is increasingly probable as we approach the faint limit, the flattening of the $\log N - \log S$ relationship as we go to fainter fluxes mimics the effects of a space distribution that is decreasing with r. In a survey where the detection limit is not well defined globally (for example, where it changes somewhat from one field to the next on account of seeing differences), simple tabulation of the source counts without any suitable correction will lead to misleading conclusions. Tests for sample completeness are usually not very rigorous. Often samples are considered to be complete (a) if the integrated surface density of detected QSOs down to some limiting flux or apparent magnitude is at least as large as in previous surveys, (b) if comparison of the final list of confirmed QSOs with a list of known QSOs in the same field reveals that few, if any, previously known QSOs were overlooked by the current survey, or (c) the slope of the $\log N - \log S$ relationship is consistent with the results of previous surveys. Quantitative comparison of various surveys is especially valuable, and for this reason, it is extremely useful to have QSO surveys with at least some overlapping fields or in areas of the sky, such as SA 57,[†] that have been well studied by a number of different techniques.

10.1.3 The Eddington Bias

Before proceeding further, we need to be aware of a potentially important effect, the Eddington bias (Eddington 1913), by which random errors in magnitude measurements can systematically alter the QSO number counts that we obtain. To quantify this effect, we use the *differential* QSO number count per magnitude per unit solid angle $A(m)$, i.e., the surface density of QSOs in the magnitude range $m - \delta m$ to $m + \delta m$ per square degree. The *cumulative* number count $N(m)$ is obviously

$$N(m) = \int_{-\infty}^{m} A(m)\,dm, \tag{10.11}$$

so that $dN(m)/dm = A(m)$.[‡] Any QSO magnitude we measure is subject to random errors, and these can increase or decrease the observed value m relative to the true value m'. Because there are more sources at true apparent magnitude $m' + \Delta m'$ than at $m' - \Delta m'$, distributed errors will tend to *increase* $A_{\text{obs}}(m)$ relative to the true value $A(m')$. This is the nature of the Eddington bias, as we will now show. (A similar derivation is carried out by Mihalas and Binney 1981, pp. 219ff.)

We assume that the random measurement errors can be described by a Gaussian distribution of width σ in $m - m'$. The probability of measuring a magnitude m for a source of true apparent brightness m' is thus given by the normalized probability distribution

$$P(m', m) = \frac{1}{(2\pi\sigma^2)^{1/2}} \exp\left[-(m - m')^2/2\sigma^2\right]. \tag{10.12}$$

The observed value of $A_{\text{obs}}(m')$ is then given by the true distribution $A(m)$ convolved with this probability distribution, i.e.,

$$A_{\text{obs}}(m') = \frac{1}{(2\pi\sigma^2)^{1/2}} \int_{-\infty}^{\infty} A(m)\,e^{-(m-m')^2/2\sigma^2}\,dm. \tag{10.13}$$

This equation can be inverted to solve for $A(m)$ by expanding $A(m)$ about m', i.e.,

$$A(m) = A(m') + (m - m')\frac{dA(m')}{dm} + \frac{(m - m')^2}{2}\frac{d^2 A(m')}{dm} + \cdots. \tag{10.14}$$

[†] The 'Selected Areas' (SA) are those which were originally designated for statistical study of the space density of stars by Kapteyn (see Blaauw and Elvius 1965). SA 57 is a well-studied field near the north Galactic pole.

[‡] Comparison with eq. (10.4) shows that $A(m) = d\mathcal{N}(m)/d\Omega$.

We can now insert eq. (10.14) into eq. (10.13), which yields

$$
\begin{aligned}
A_{\text{obs}}(m') &= \frac{1}{(2\pi\sigma^2)^{1/2}} \int_{-\infty}^{\infty} A(m')\, e^{-(m'-m)^2/2\sigma^2}\, dm \\
&+ \frac{1}{(2\pi\sigma^2)^{1/2}} \int_{-\infty}^{\infty} \frac{dA(m')}{dm} (m-m')\, e^{-(m'-m)^2/2\sigma^2}\, dm \\
&+ \frac{1}{(2\pi\sigma^2)^{1/2}} \int_{-\infty}^{\infty} \frac{d^2A(m')}{dm^2} \frac{(m-m')^2}{2}\, e^{-(m'-m)^2/2\sigma^2}\, dm \\
&= A(m') + 0 + \frac{\sigma^2}{2} \frac{d^2A(m')}{dm^2},
\end{aligned}
\tag{10.15}
$$

where the second term vanishes because it is an odd integral – in other words, to first order, the number of sources we measured as being too bright is canceled by the number that we measured as being too faint. As long as σ is small, the non-zero higher-order terms (of order σ^4 and higher) can be ignored, and we can also approximate $d^2A(m')/dm^2 \approx d^2A_{\text{obs}}(m')/dm^2$, which gives

$$
A(m) \approx A_{\text{obs}}(m) - \frac{\sigma^2}{2} \frac{d^2A_{\text{obs}}(m)}{dm^2}.
\tag{10.16}
$$

The second term is thus a small correction to the observed differential counts $A_{\text{obs}}(m)$, which will otherwise be too large.

This can be put into a more convenient form by first generalizing eq. (10.8) to the form

$$
N(m) \propto C\, 10^{\kappa m},
\tag{10.17}
$$

where C is a constant and we will consider κ to be constant or a slowly varying function of m; in a constant-density Euclidean space $\kappa = 0.6$ (eq. 10.8). We can then write

$$
A(m) = \frac{dN(m)}{dm} = C\kappa\,(\ln 10)\, 10^{\kappa m},
\tag{10.18}
$$

so that $\log A(m) = C' + \kappa m$, where C' is also a constant. Noting that

$$
\frac{d\log A}{dm} = \frac{\log e}{A} \frac{dA}{dm},
\tag{10.19}
$$

we obtain

$$
\frac{d^2A}{dm^2} = \frac{dA}{dm} \frac{1}{\log e} \frac{d\log A}{dm} = A \left(\frac{1}{\log e} \frac{d\log A}{dm} \right)^2,
\tag{10.20}
$$

and thus eq. (10.16) can be written

$$
A(m) \approx A_{\text{obs}}(m) \left[1 - \frac{1}{2} \left(\frac{\sigma\kappa}{\log e} \right)^2 \right].
\tag{10.21}
$$

For a sample characterized by typical magnitude errors $\sigma \approx 0.1\,\mathrm{mag}$ and $\kappa \approx 0.6$, eq. (10.21) shows that the observed surface-density counts must be reduced by only 1% and therefore the correction can be neglected. However, if we consider the case $\sigma \approx 0.3\,\mathrm{mag}$ and $\kappa \approx 0.8$, both of which are realistic estimates in some cases, the correction factor is $\sim 15\%$.

10.1.4 Survey Selection Functions

In real situations, obtaining a complete sample of QSOs to any reasonable flux limit is an impossibility (Weedman 1986, Hewett and Foltz 1994). In addition to the dependence on luminosity, redshift, and SED, the probability of detecting a particular source depends on several other parameters, such as:

(1) **Variability:** Because the luminosities vary with time, sources near the detection limit might at various times be above or below the detection threshold. If the variations can be plausibly regarded as Gaussian-distributed in magnitude about some mean value, then the formalism for correcting the sample counts is identical to the correction for the Eddington bias.

(2) **Emission-line equivalent widths:** The contrast between lines and continuum can vary significantly, even among QSOs with similar luminosities and SEDs. Surveys that depend in one way or another on line contrast may find the strong-lined objects but miss those with weaker lines. If we can quantify the sensitivity of a survey to emission-line equivalent width and we have some idea of the true distribution of equivalent widths, we can make a suitable statistical correction to the measured surface densities.

(3) **Absorption lines:** As we will see in Chapter 12, spectra of high-redshift QSOs tend to show the absorption signatures of unrelated objects along the line of sight. This can significantly alter the appearance of the QSO spectrum, particularly at wavelengths shortward of the Lyα emission line where the density of strong absorption lines per unit wavelength interval is quite high, thus suppressing the total continuum flux. This can change the probability of detecting such a QSO relative to an identical, but unabsorbed, source.

(4) **Internal absorption:** There is evidence that some AGN spectra are reddened and absorbed by dust that presumably resides in the host galaxies or possibly in the QSO line-emitting regions (MacAlpine 1985). The spectra of some QSOs might be absorbed by dust in their host galaxies, particularly if they are seen close to edge-on (De Zotti and Gaskell 1985). Some low-redshift AGNs appear to be reddened by more than $E(B - V) > 0.1\,\mathrm{mag}$, which corresponds to extinction in excess of 0.8 mag at a rest wavelength of 1500 Å. This would greatly reduce the probability of detecting $z \approx 2$ QSOs with comparable amounts of dust. In some cases, intrinsically luminous sources might be missed altogether if the observer is unfavorably located, for example, in the plane of an obscuring torus (Chapter 7). Indeed, Webster *et al.* (1995) have argued that many, and perhaps even *most*,

AGNs have been missed in traditional optical surveys because their light is so heavily absorbed by internal dust.

A recent practice is to try to deal with survey incompleteness in a realistic fashion by quantitative assessment of the survey selection function (e.g., Warren, Hewett, and Osmer 1994). The basic idea is to assume that effects such as the four listed above are stochastically operative on an ensemble of otherwise identical QSOs in a given comoving volume. For a given luminosity, redshift, and intrinsic SED, we carry out Monte Carlo simulations that represent the range of these various effects on the original spectrum. A single Monte Carlo realization produces from the original spectrum a model spectrum that is altered by some combination of the particular modifying effects, and the model spectrum is used to produce simulated data by convolving the model spectrum with the wavelength-dependent survey sensitivity (e.g., by convolving the model spectrum with the filter bandpasses to obtain simulated magnitudes in each bandpass). The simulated data are then passed through the selection algorithms to determine whether or not the model QSO would have been identified as a QSO candidate or rejected as a probable contaminant. By carrying out an extensive array of such simulations in which the various effects that can alter the appearance of the QSO vary from simulation to simulation, we can determine what fraction of all QSOs with a given luminosity, redshift, and SED would be successfully identified by the survey, and then correct the measured surface density $d\mathcal{N}(F, z)/d\Omega$ to account for the QSOs that we would be unable to detect.

10.2   Color Selection

Because AGN SEDs differ so significantly from those of stars, color selection has proven to be a highly effective way of distinguishing between QSOs and stars. The principal difficulty with color-based selection is that QSO colors have a very strong redshift dependence, so the probability of detecting a given QSO is a function of z. Understanding the dependence of the survey selection function on redshift is thus critical in order to distinguish real changes in the QSO space density from survey biases.

10.2.1 K-Corrections

The redshift dependence of the magnitude of any object in a given wavelength band is expressed in terms of a 'K-correction' (Oke and Sandage 1968), which is defined as

$$m_{\text{intrinsic}} = m_{\text{observed}} - K(z), \tag{10.22}$$

or equivalently expressed in fluxes,

$$F_{\text{intrinsic}} = F_{\text{observed}} \, 10^{0.4K(z)}. \tag{10.23}$$

In general, $K > 0$, since the SEDs of most objects decrease at shorter wavelengths, so a Doppler shift decreases the apparent brightness of a receding source. The K-correction is formally defined as

$$K(z) = K_1 + K_2, \tag{10.24}$$

where the two terms are

$$K_1 = 2.5 \log(1 + z) \tag{10.25}$$

and

$$K_2 = 2.5 \log \left[\int_0^\infty F(\lambda) S(\lambda) d\lambda \Big/ \int_0^\infty F(\lambda/[1+z]) S(\lambda) d\lambda \right]. \tag{10.26}$$

Here $F(\lambda)$ is the SED of the object observed (galaxy or quasar) and $S(\lambda)$ is the filter response function. The first term (K_1) in the K-correction is due to the narrower width of the filter in the observer's frame; a spectral region of width $\delta\lambda_0$ is stretched to width $\delta\lambda_0(1 + z)$ in the observer's frame, which is equivalent to apparent contraction of the observer's filter by the same factor. The integrals in the second term (K_2) are the convolution of the spectrum with the filter response in the observer's frame (in the denominator), normalized by the value at rest (in the numerator).

It is instructive to consider the K-correction for a simple power-law SED $F_\nu = C\nu^{-\alpha}$, since this provides a good low-order approximation to a real AGN continuum. We can rewrite this in wavelength-dependent units as $F_\lambda = C'\lambda^{\alpha-2}$ (see §9.4.1). Thus

$$F_\lambda\left(\frac{\lambda}{1+z}\right) = C'\left(\frac{\lambda}{1+z}\right)^{\alpha-2} = F_\lambda(\lambda)(1+z)^{2-\alpha}, \tag{10.27}$$

so the second term in eq. (10.24) becomes

$$\begin{aligned}
K_2 &= 2.5 \log \left[\int_0^\infty F(\lambda) S(\lambda) d\lambda \Big/ \int_0^\infty F(\lambda/[1+z]) S(\lambda) d\lambda \right] \\
&= 2.5 \log \left[\int_0^\infty F(\lambda) S(\lambda) d\lambda \Big/ (1+z)^{2-\alpha} \int_0^\infty F(\lambda) S(\lambda) d\lambda \right] \\
&= 2.5 \log(1+z)^{\alpha-2} \\
&= 2.5(\alpha - 2) \log(1 + z).
\end{aligned} \tag{10.28}$$

Equation (10.24) thus becomes

$$K(z) = 2.5(\alpha - 1)\log(1 + z), \tag{10.29}$$

so

$$m_{\text{intrinsic}} = m_{\text{observed}} - 2.5(\alpha - 1)\log(1 + z), \tag{10.30}$$

or

$$F_{\text{intrinsic}} = F_{\text{observed}}(1 + z)^{\alpha-1}. \tag{10.31}$$

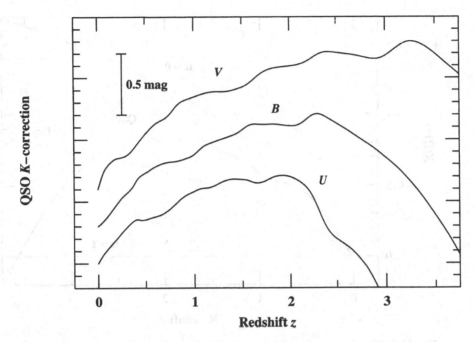

Fig. 10.1. Nominal K-corrections (eq. 10.24) for Johnson U, B, and V bands as a function of QSO redshift. These K-corrections, in magnitudes, are computed from the mean QSO spectrum shown in Fig. 2.2. The zero points of the curves have been shifted to avoid overlap; the K-correction in each band is zero at $z = 0$. The K-corrections for QSOs are *negative* for $z > 0$.

For $\alpha \approx 1$, the K-correction vanishes. For flatter spectra ($\alpha < 1$), the K-correction is *negative*, as such objects are quite blue (compared to stars), and the Doppler shift brings a brighter part of the spectrum into the observed waveband.

The K-correction is less simple when the emission lines are properly taken into account. The K-correction is a complicated function of z, since it depends on which emission lines are redshifted into the filter. The magnitude of the K-correction depends on the equivalent widths of the emission lines and obviously the largest corrections occur when the strongest lines, Lyα and C IV in particular, appear in a given waveband. This is shown for the Johnson U, B, and V bands in Fig. 10.1.

Nominal K-corrections can be computed by convolving a typical QSO spectrum at different redshifts with the filter response. Such a computation is shown graphically in Fig. 10.2.

10.2.2 Color-Induced Biases

Much of the importance in understanding the K-corrections for QSOs derives from the fact that the presence or absence of the strong emission lines in bandpasses used for QSO searches affects the probability of detection. For example, the Lyα line is

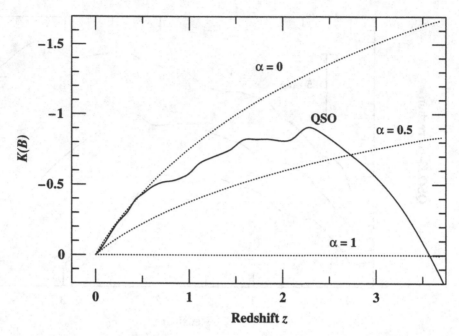

Fig. 10.2. The K-correction for the Johnson B band as a function of QSO redshift, as in Fig. 10.1. Also shown are the B-band K-corrections for power-law spectra, $F_\nu \propto \nu^{-\alpha}$; as described in the text, the K-correction vanishes for $\alpha = 1$. The structure in the QSO K-correction is due to strong emission lines moving through the bandpass as redshift increases. The K-correction has its largest value when Lyα enters the band. The values at $z \gtrsim 3$ are not reliable. After a similar diagram by Hewett (1992).

centered in the U, B, and V filters at redshifts $z \approx 2$, $z \approx 2.6$, and $z \approx 3.5$, respectively, thus enhancing the probability of detecting QSOs at these redshifts on account of the unusual colors they will have. Figure 10.1 shows that when the Lyα line enters the U band, the apparent U magnitude of the QSO is enhanced by ~ 0.4 mag relative to what would be measured for the same QSO with no Lyα line. This would bring QSOs that are nominally as much as 0.4 mag fainter than our detection limit into our sample at this redshift. This would enhance the number of QSOs at this redshift by a factor (eq. 10.8) $10^{0.6 \times 0.4} \approx 1.7$. In fact, this rather underestimates the effect because for $m \lesssim 19$ mag, the observed value of $d \log N(m)/dm \approx 0.8$ (Hartwick and Schade 1990). We thus expect to find significantly more QSOs at redshifts where one of the strong emission lines is in the bandpass on account of the combination of the strength of the broad emission lines and the steepness of the QSO luminosity function. This is a serious selection effect which can alter the apparent space density of QSOs as a function of redshift. Indeed, there were claims of 'periodicities' in QSO counts as a function of redshift, and these apparent fluctuations in the number counts were due to color-based selection criteria. This is demonstrated very clearly in Fig. 10.3.

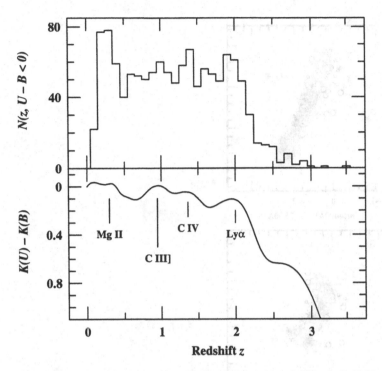

Fig. 10.3. The top panel shows the distribution in redshift of all 1239 QSOs in the Hewitt and Burbidge (1993) catalog for which U and B magnitudes are available and for which $U - B < 0$. Nearly all of these sources were identified by this 'UV excess'. The lower panel shows the *difference* between the U and B K-corrections, which is a measure of the relative sensitivity to $U - B$ color selection at different redshifts. The relative peaks in the K-correction occur when the strongest emission lines are centered in the U bandpass, as labeled in the lower panel. The distribution in the top panel shows peaks at these same redshifts on account of the increased sensitivity of the $U - B$ technique at redshifts where the U brightness is enhanced by emission lines in the band. After a similar diagram by Kjærgaard (1978).

10.2.3 Color-Selected QSOs

As we have noted in several places, color selection of QSOs has traditionally meant by UV excess. Probably the best-known large-scale UV-excess survey is the Palomar Bright Quasar Survey (BQS), which yielded the Palomar–Green (PG) QSOs (Schmidt and Green 1983). The candidate objects in this survey were selected from digitized photographic plates according to the criterion $U - B < -0.44$. The limiting magnitude of survey plates varied somewhat, but was nominally around $B_{\text{lim}} \approx 16\,\text{mag}$.

The UV-excess method works quite well up to redshifts $z \approx 2$. The most important contaminant in these surveys is white dwarfs, which are also distinguished by $U - B \lesssim 0$.

At redshifts $z > 2$, $U - B$ becomes less effective in isolating QSOs, and in fact becomes very ineffective for finding the highest-z sources once the Lyα emission line is

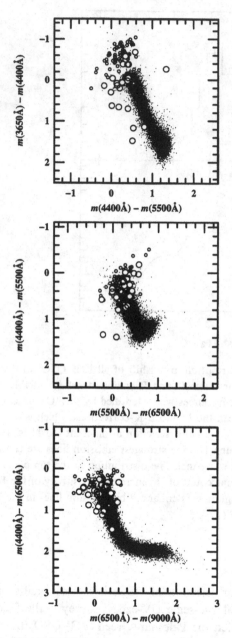

Fig. 10.4. These are two-color diagrams for the brighter ($16.00 \leq m(6500\,\text{Å}) \leq 18.50$) stellar (point-like) objects on south Galactic pole survey plates. The black dots show the distribution of stars measured from the survey plates. The large open circles represent quasars with $z \geq 2.2$, and the small open circles represent quasars with $z < 2.2$. Data from Warren *et al.* (1991), provided by S. J. Warren.

redshifted beyond the U band and all that remains is the highly absorbed continuum (Chapter 12).

One of the most important lessons of this chapter is that the overall efficiency with which QSOs can be identified on the basis of a single color, such as $U - B$, varies significantly with redshift. It is therefore desirable to base a survey on a larger number of colors that cover the full range of the easily observed optical/near-IR spectrum; whereas at certain redshifts a QSO SED might be hard to distinguish from stars, a QSO is extremely unlikely to be indistinguishable from a star in multiple colors. The 'UK Schmidt Survey' (Warren *et al.* 1991), which covers more than $40 \deg^2$ in six separate bandpasses, provides an example of the multicolor approach. The photographic data from this survey were digitized with the Cambridge Automatic Plate Measuring (APM) facility. Multiple colors were computed from the digitized data for all the non-extended objects. Stars are found to be localized in the multidimensional color space, and QSO candidates were selected from the points which were significantly outside the stellar locus in at least one color (Fig. 10.4).

10.3 Slitless Spectroscopy

While the efficiency of UV-excess color selection decreases dramatically at $z \gtrsim 2$, slitless spectroscopy is particularly effective at isolating high-redshift QSOs, primarily because the strong Lyα and C IV emission lines are redshifted into the optical spectrum where they can be easily observed from the ground. As with multicolor selection, the dramatic differences between QSO SEDs and those of stars are used to isolate the QSOs, and the efficacy of the method derives from the ability to 'multiplex', or record at one time the spectra of many objects, only a few of which will turn out to be of interest. Compared to color selection, the efficiency of QSO detection varies relatively slowly with z, basically because a larger continuous portion of the spectrum is recorded by using low-dispersion spectroscopy. While the exposure times for individual observations are longer than for narrow-band imaging, the number of exposures per field is smaller, and thus the overall efficiency in units of telescope time is about the same for the two methods (Pritchet and Hartwick 1987).

The dispersing element used is sometimes an objective prism, but more often a 'grism' (a prism with one side ruled as a grating), which has several advantages, including better throughput and production of a 'zero-order' image which allows one to measure the displacement of an emission line from the zero-order point and thus determine the redshift at least approximately.

As with other QSO survey methods, there are inherent limitations and selection biases that need to be understood. The basic problems with slitless spectroscopy are:

(1) The limiting flux for slitless spectroscopy is significantly higher than for broad-band imaging. The reason for this is that the flux from a faint source is dispersed by wavelength over many pixels on the detector. The critical factor, however, is

that the sky brightness in any pixel is *not* correspondingly diminished because each pixel receives the sky signal at all wavelengths that reach the detector, although not from the same area of the sky. Thus the effective signal-to-noise ratio relative to a broad-band image is reduced.

(2) The limiting magnitude of the survey material cannot be reliably determined from the spectroscopic data – identification of a QSO depends often on emission-line identification, and the continuum flux of a particular faint QSO might lie well below the nominal detection limit without actually precluding detection of the source. A way around this difficulty is to obtain broad-band images of the field to define *a priori* a magnitude-limited sample in the field (e.g., Foltz *et al.* 1987).

(3) Slitless spectra have limited wavelength information, so sometimes there are few (or no) lines in the spectra. This is a particular problem at $z \approx 1$, when only Mg II $\lambda 2798$ is redshifted into the optical spectrum.

(4) There is a clear bias in such surveys towards objects with the strongest lines relative to the continuum (Peterson 1988). With photographic slitless spectroscopy, weak-lined objects are a particular problem because of their very low contrast. Statistically, this operates slightly against detection of higher-luminosity QSOs on account of the Baldwin effect (§5.8.2). The bias between line-selected and continuum-selected objects can be quantified, however (e.g., Schmidt, Schneider, and Gunn 1995).

(5) Slitless spectra are sometime confused in crowded fields. Field-crowding problems can be significant even at the lowest resolution at which one can hope to see emission lines, and the spectra of adjacent objects tend to overlap. This can lead to errors unless great care is taken (for example, by multiple exposures with the dispersion in different directions). However, it can be argued (Hewett and Foltz 1994) that these problems can be modeled in the survey selection function, and that the telescope time is more effectively used by increasing the area of sky surveyed than by obtaining redundant exposures of fields for an incremental improvement in completeness.

As in the case of color surveys, spectroscopic surveys were originally carried out with photographic detectors and the data were inspected visually. Even when the detection limit and overall efficiency were poorly known, these surveys led to important results, such as Osmer's (1982) recognition that the space density of very high-redshift ($z \gtrsim 3.5$) QSOs must be extremely low. Also, as with the color-based surveys, the basic data are now either digitized from photographic plates or obtained with CCDs, and thus sophisticated candidate-selection algorithms can be employed (e.g., Hewett *et al.* 1985). An example of such a program is the Large Bright Quasar Survey (LBQS) which has found over 1000 QSOs with $B \lesssim 18.7$ mag (Hewett, Foltz, and Chaffee 1995).

Table 10.1

Surface Density of QSOs

B (mag)	$\log N(B, z < 2.2)$ (deg^{-2})	$\log N(B, z > 2.2)$ (deg^{-2})
13.0	−4.0	
13.5	−3.7	
14.0	−3.55	
14.5	−3.24	
15.0	−2.77	
15.5	−2.39	
16.0	−1.99	
16.5	−1.70	
17.0	−1.23	
17.5	−0.73	−1.8
18.0	−0.23	−1.4
18.5	0.12	−0.69
19.0	0.63	−0.12
19.5	0.93	0.12
20.0	1.21	0.38
20.5	1.40	0.62
21.0	1.52	0.88
21.5	1.73	1.1
22.0	1.87	1.3
22.5	2.11	1.5

10.4   Other Selection Methods

While slitless spectroscopy and color selection are probably the most efficient means of isolating QSO candidates, other methods deserve at least brief mention because they provide important checks on the principal techniques.

QSOs typically vary in brightness by a few tenths of a magnitude on time scales of a year or so, and this distinguishes them from most stars. Identification of QSO candidates on the basis of variability has proven to be quite effective when based on a sufficiently large number of observations spaced over many years (Usher and Mitchell 1978, Hawkins 1986, Hook *et al.* 1994). There is statistically a $(1 + z)$ bias against selection of high-redshift objects on the the basis of variability on account of time-dilation effects, and the inverse correlation between luminosity and amplitude of variability introduces a bias against selection of high-luminosity sources; again, in principle, these effects can be modeled and statistical corrections applied (Hook *et al.* 1994).

QSOs can be distinguished from nearby Galactic objects, notably white dwarfs, from their lack of proper motion (Sandage and Luyten 1967, Kron and Chiu 1981, Koo, Kron, and Cudworth 1986). While this is by itself not a highly effective method of isolating QSOs, it can be useful in combination with other criteria (e.g., variability, colors) and in verifying that previously unsuspected classes of QSOs have not been overlooked. Surveys based on observations at other wavelengths, such as the X-ray *Einstein* Medium-Sensitivity Survey (Gioia *et al.* 1990, Stocke *et al.* 1991), provide a similarly important check on the conclusions drawn from the optically selected samples.

10.5   The Surface Density of QSOs

The cumulative distribution of QSOs $N(B)$ has been constructed from a variety of sources by Hartwick and Schade (1990), and their results are given in Table 10.1. Inspection of this table reveals that the slope $d \log N(B)/dB$ has a value ~ 0.8 for $B \lesssim 19$ mag, as mentioned earlier, which is far steeper than the Euclidean prediction of 0.6. At the faintest magnitudes ($B \gtrsim 20$ mag), however, the slope becomes quite shallow ($d \log N(B)/dB \approx 0.35$). We will see in the next chapter what this implies about the evolution of the QSO population.

11 The Quasar Luminosity Function and Evolution

As a direct result of advances in the QSO survey techniques described in Chapter 10, the number of QSOs known as of the early 1990s was of order 10^4 (Véron-Cetty and Véron 1991, Hewitt and Burbidge 1993). It is therefore possible to explore the distribution of QSOs as a function of z, to compute their comoving space density, and to determine how the AGN population evolves with time.

11.1 Simple Tests for Evolution

Determination of the space density and luminosity function of QSOs is a difficult undertaking, and as seen in Chapter 10, the results are sensitive to many different types of selection effects. Despite these many difficulties, it was clear even during the first decade of QSO research that the comoving density of QSOs varies strongly with redshift, with an especially large number of QSOs at $z \approx 2$. However, it is also at about this redshift that UV-excess techniques are most sensitive, since this is where the Lyα emission line falls in the U band, so originally it was not entirely clear to what extent the inferred high densities of QSOs at $z \approx 2$ were due to selection effects rather than a real peak in the comoving space density. For this reason we begin this chapter with a discussion of simple tests for changes in space density as a function of z, or equivalently, lookback time.

11.1.1 The Log N – Log S Test in a Non-Euclidean Universe

The classic test for a constant space density of sources is the log N – log S test that was introduced in Chapter 10. The great advantage of the log N – log S test is that it does not require knowing the distance to individual sources. For this reason it was especially important in analyzing the distribution of radio sources as it was not necessary also to obtain optical spectra to measure their redshifts.

Luminous radio sources (including quasars) often lie at large cosmological distances (e.g., see Spinrad *et al.* 1985), and thus it is necessary to consider the effect of a non-Euclidean Universe on source counts. We suppose then that the radio spectrum can be modeled as a power law, i.e., with specific luminosity $L_\nu \propto \nu^{-\alpha}$, and for radio galaxies and quasars $\langle \alpha \rangle \approx 0.7$. The flux measured at some observed frequency ν_0 was emitted at the source at frequency $\nu_1 = \nu_0(1 + z)$. From eq. (9.98) or Table 9.1, the flux in the observer's frame is related to that in the source frame by

$$S_\nu(\nu_0) = \frac{S_\nu(\nu_1)}{(1+z)},$$ (11.1)

and the flux and luminosity are related by (cf. eq. 9.102)

$$S_\nu(\nu_1) = \frac{L_\nu(\nu_1)}{4\pi(a_0 r_1)^2}.$$ (11.2)

By combining these and using the equation for the luminosity distance d_L (eq. 9.104), we have

$$S_\nu(\nu_0) = \frac{S_\nu(\nu_1)}{(1+z)} = \frac{L_\nu(\nu_1)(1+z)}{4\pi d_L^2}.$$ (11.3)

Since the spectrum is a power law,

$$\frac{L_\nu(\nu_1)}{L_\nu(\nu_0)} = \left(\frac{\nu_1}{\nu_0}\right)^{-\alpha} = (1+z)^{-\alpha}$$ (11.4)

and thus

$$S_\nu(\nu_0) = \frac{L_\nu(\nu_0)(1+z)^{1-\alpha}}{4\pi d_L^2}.$$ (11.5)

This same equation would hold for a redshifted object in Euclidean space if we replace the luminosity distance d_L with the distance r.

Generalizing the development of Chapter 10, we can write the surface density of sources at redshift z as

$$d\mathcal{N}(z) = n(z)\, dV(z) = n(z) a_0^2 r^2 dr\, d\Omega.$$ (11.6)

The calculation of the volume element $dV(z)$ is fairly straightforward. Using eq. (9.90) and eq. (9.104), we have

$$a_0 r = \frac{c}{H_0 q_0^2} \left[\frac{z q_0 + (q_0 - 1)(-1 + \sqrt{2q_0 z + 1})}{1+z} \right] = \frac{d_L}{1+z},$$ (11.7)

and by using eq. (9.63) we can write

$$dr = \frac{c\, dz}{H(z)} = \frac{c\, dz}{H_0(1+z)(1+2q_0 z)^{1/2}}.$$ (11.8)

Combining these gives

$$dV(z) = a_0^2 r^2 dr\, d\Omega = \frac{d_L^2 c\, dz\, d\Omega}{H_0(1+z)^3(1+2q_0 z)^{1/2}}.$$ (11.9)

The assumption that we will test observationally is that the comoving density of radio sources is constant with z, i.e., $n(z) = n_0(1+z)^3$. The surface density of sources at z is thus $dN(z) = d\mathcal{N}(z)/d\Omega$ and therefore

$$\frac{dN(z)}{dz} = \frac{d_L^2 n_0 c}{H_0(1+2q_0 z)^{1/2}}.$$ (11.10)

From eq. (11.5), we can write

$$
\begin{aligned}
\frac{dS_v}{dz} &= \frac{L_v}{4\pi} \frac{d}{dz} \left(d_L^{-2} (1+z)^{1-\alpha} \right) \\
&= \frac{L_v}{4\pi} \left(-2 d_L^{-3} (1+z)^{1-\alpha} \frac{d(d_L)}{dz} + (1-\alpha) d_L^{-2} (1+z)^{-\alpha} \right) \\
&= \frac{-L_v (1+z)^{1-\alpha}}{4\pi d_L^2} \left(\frac{2}{d_L} \frac{d(d_L)}{dz} + \frac{(\alpha-1)}{(1+z)} \right).
\end{aligned}
\tag{11.11}
$$

We can combine eqs. (11.10) and (11.11) to obtain the differential counts per unit flux dN/dS. While this expression is obviously fairly complicated, we can simplify things a great deal by comparing this quantity to what we would expect in the Euclidean case (Longair 1978). Starting with eq. (10.5) and noting from eq. (11.5) that $L_v/4\pi S_v = d_L^2 (1+z)^{\alpha-1}$, we have for the Euclidean case

$$
\begin{aligned}
\frac{dN_E}{dS} &= \frac{-n_0}{2} \left(\frac{L_v}{4\pi S_v} \right)^{3/2} S_v^{-1} \quad (\propto S^{-5/2}) \\
&= \frac{-n_0}{2} \left(d_L^2 (1+z)^{\alpha-1} \right)^{3/2} \left(\frac{4\pi d_L^2 (1+z)^{\alpha-1}}{L_v} \right) \\
&= \frac{-2\pi n_0}{L_v} d_L^5 (1+z)^{5(\alpha-1)/2}.
\end{aligned}
\tag{11.12}
$$

By combining eqs. (11.10), (11.11), and (11.12), we obtain

$$
\frac{dN/dS}{dN_E/dS} = \frac{2c (1+z)^{-3(\alpha-1)/2}}{d_L H_0 (1+2q_0 z)^{1/2}} \left[\frac{2}{d_L} \frac{d(d_L)}{dz} + \frac{(\alpha-1)}{(1+z)} \right]^{-1}.
\tag{11.13}
$$

While eq. (11.13) is also rather unwieldy, it simplifies considerably for the Einstein–de Sitter case where

$$
d_L(q_0 = 1/2) = \frac{2c}{H_0} \left((1+z) - (1+z)^{1/2} \right)
\tag{11.14}
$$

and

$$
\left. \frac{d(d_L)}{dz} \right|_{q_0=1/2} = \frac{2c}{H_0} \left(1 - \frac{1}{2(1+z)^{1/2}} \right),
\tag{11.15}
$$

so we obtain

$$
\Theta \equiv \left. \frac{dN/dS}{dN_E/dS} \right|_{q_0=1/2} = \frac{(1+z)^{-3(\alpha-1)/2}}{(1+\alpha)(1+z)^{1/2} - \alpha}.
\tag{11.16}
$$

For the mean spectral index $\langle \alpha \rangle = 0.7$, the function Θ decreases monotonically from unity at $z = 0$ to values of 0.80 and 0.73 at $z = 1$ and $z = 2$, respectively. For reasonable combinations of q_0 and α, $\Theta \lesssim 1$, which implies that the slope of the log N – log S relationship is expected to be shallower (i.e., closer to zero) than the Euclidean value of $-3/2$.

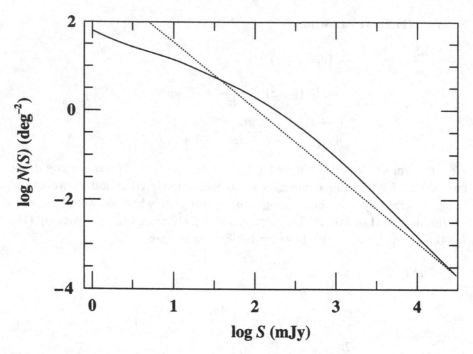

Fig. 11.1. A plot of log N – log S for radio sources observed at a frequency of 1.4 GHz (Condon 1984). The dotted line is the Euclidean prediction $N(S) \propto S^{-3/2}$. At fluxes higher than ~ 1 Jy, the observed slope is *steeper* than this prediction, i.e., as we go to fainter fluxes, we find more sources than we expect on the basis of bright source counts. In the Einstein–de Sitter case, we would expect the slope to be *shallower* than in the Euclidean case. We conclude that the there are more radio sources at large distances (or lookback times) than there are locally. The observed log N – log S relationship becomes flatter at $F_\nu(1.4\,\text{GHz}) \lesssim 100$ mJy, consistent with the prediction of the non-Euclidean case. Strong evolution is thus inferred only for the most luminous radio sources. Above 10 mJy, the population of radio sources is dominated by quasars and giant ellipticals.

11.1.2 Results of the Log N – Log S Test

Results of the log N – log S test for strong radio sources at 1.4 GHz are shown in Fig. 11.1. Despite the prediction that $d \log N(S)/d \log S \geq -1.5$, at the high-flux end the slope is clearly steeper than this, i.e., $d \log N(S)/d \log S \approx -1.8$. This is not expected for a constant comoving density in any cosmological model we have considered, and we must therefore conclude that the population of radio sources is evolving with time, in the sense that there were more sources per unit comoving volume in the past than there are now.

The optical log $N(m)$ – log m relationship is similarly expected to have a slope $d \log N(m)/dm \leq 0.6$ in the non-Euclidean case. Again the optical sources counts (Table 10.1) show that contrary to the assumption of constant comoving density the slope significantly exceeds 0.6 for $m_B \lesssim 19$ mag.

11.1.3 The Luminosity–Volume Test

In an attempt to circumvent the limitations of the $\log N - \log S$ test, Schmidt (1968) devised a test for evolution that is not sensitive to the completeness of the sample and is moreover especially useful when there are multiple observational constraints, such as both radio and optical detection limits. This is known as the 'luminosity–volume test', or more commonly by the name of the statistic used, i.e., the 'V/V_{max} test'.

Suppose that we detect a source at some distance r. In Euclidean space, the observed flux is $F = L/4\pi r^2$, with $F \geq S$, the detection limit. This same source in principle could have been detected if it were as far away as $r_{max} = (L/4\pi S)^{1/2}$. This maximum distance r_{max} encloses a spherical volume around the observer

$$V_{max} = \frac{4\pi r_{max}^3}{3} \tag{11.17}$$

within which a source of luminosity L could in principle have been detected. Now if we consider many such sources, we expect that half of the sources will be found in the inner half of this volume, and the rest will be found in the outer half, provided that the density of such objects is uniform. Thus we can form a useful statistic V/V_{max}, where V is the spherical volume on whose surface the object is actually located; we expect that the average value of V/V_{max} will be 0.5 if the objects are distributed uniformly in space. This can be demonstrated by computing the mean value of V for a constant density sample,

$$
\begin{aligned}
\langle V \rangle &= \int_\Omega \int_0^{r_{max}} (4\pi r^3/3)\, n(r) r^2 dr\, d\Omega \bigg/ \int_\Omega \int_0^{r_{max}} n(r) r^2 dr\, d\Omega \\
&= \frac{4\pi}{3} n_0 \int_0^{r_{max}} r^5 dr \bigg/ n_0 \int_0^{r_{max}} r^2 dr \\
&= \frac{4\pi}{3} \frac{r_{max}^6/6}{r_{max}^3/3} = \frac{4\pi}{3} \frac{r_{max}^3}{2},
\end{aligned}
\tag{11.18}
$$

so for a uniform density $\langle V/V_{max} \rangle = 1/2$. We can similarly compute a formal uncertainty on the measured value of $\langle V/V_{max} \rangle$ for a sample of N objects from

$$\sigma_{V/V_{max}} = \left[\frac{1}{N} \left(\left\langle \frac{V^2}{V_{max}^2} \right\rangle - \left\langle \frac{V}{V_{max}} \right\rangle^2 \right) \right]^{1/2}. \tag{11.19}$$

In analogy with eq. (11.18), $\langle V^2 \rangle$ can be computed in a straightforward way:

$$
\begin{aligned}
\langle V^2 \rangle &= \int_\Omega \int_0^{r_{max}} (4\pi r^3/3)^2\, n(r) r^2 dr\, d\Omega \bigg/ \int_\Omega \int_0^{r_{max}} n(r) r^2 dr\, d\Omega \\
&= \left(\frac{4\pi}{3}\right)^2 n_0 \int_0^{r_{max}} r^8 dr \bigg/ n_0 \int_0^{r_{max}} r^2 dr \\
&= \left(\frac{4\pi}{3}\right)^2 \frac{r_{max}^9/9}{r_{max}^3/3} = \frac{1}{3} \left(\frac{4\pi r_{max}^3}{3}\right)^2 = \frac{V_{max}^2}{3}.
\end{aligned}
\tag{11.20}
$$

Equation (11.19) becomes

$$\sigma_{V/V_{max}} = \frac{1}{N^{1/2}} \left[\frac{1}{3} - \left(\frac{1}{2} \right)^2 \right]^{1/2} = \left(\frac{1}{12N} \right)^{1/2}. \tag{11.21}$$

The V/V_{max} test is thus very easy to carry out: we compute the average value of r^3/r_{max}^3 for the N objects in the sample and the associated uncertainty of this mean value $\sigma_{V/V_{max}} = (12N)^{-1/2}$ to determine whether or not the mean value is significantly different from the constant-density value of 0.5.

The calculation of this statistic is only somewhat more complicated for an expanding Universe. First, the luminosity distance d_L (eq. 9.104), must be used in place of both r and r_{max}. Second, suitable K-corrections must also be applied. The way that this is usually done is to compute the absolute magnitude of the source observed to have apparent magnitude m at luminosity distance d_L (measured in Mpc),

$$M = m - K(z) - 5 \log d_L - 25. \tag{11.22}$$

We then solve for the value of z_{max} at which a source of absolute magnitude M would be at the detection limit m_{lim}. Finally, the constant-density assumption becomes one of constant density per unit *comoving volume*. A comoving volume element at coordinate r is $dV(z) = a_0^2 r^2 dr \, \delta\Omega$ and from eq. (11.9) we have

$$\frac{dV}{dz} = \frac{d_L^2 c \, dz \, \delta\Omega}{H_0 (1+z)^3 (1 + 2q_0 z)^{1/2}}. \tag{11.23}$$

For each of the QSOs in the sample ($i = 1, \ldots, N$), the volume integral

$$\begin{aligned} V_i(z') &= \int_0^{z'} \frac{dV(z)}{dz} dz \\ &= \int_0^{z'} \frac{d_L^2(z) c \, dz}{H_0 (1+z)^3 (1 + 2q_0 z)^{1/2}}. \end{aligned} \tag{11.24}$$

must be evaluated for both the QSO redshift $z' = z$ and the appropriate maximum redshift at which the QSO could have been detected $z' = z_{max}$, where we have ignored the integral over $\delta\Omega$ since it cancels in the V/V_{max} ratio. We then compute

$$\left\langle \frac{V}{V_{max}} \right\rangle = \sum_{i=1}^{N} \frac{V_i(z)}{V_i(z_{max})}. \tag{11.25}$$

A strength of the V/V_{max} test is that multiple selection criteria can be used. As an obvious example, suppose that we are obtaining optical spectra of radio sources to see if they are really QSOs (which was one of the first applications of this test). Consider the following possible outcomes:

(1) A fairly bright radio source could be seen at a large distance, and thus have a large associated value of z_{max}. However, this source could turn out to be very faint optically, so $z_{max}(\text{radio}) > z_{max}(\text{optical})$.

(2) A faint radio source which barely makes it into the candidate list might turn out to be associated with an optically bright object. If it were only slightly fainter in the radio, it would have gone undetected despite the fact that in the optical it could have been found even at much larger distances. In this case, z_{max}(optical) $> z_{max}$(radio).

Which of the two possibilities, z_{max}(radio) or z_{max}(optical), should be used in this test? The answer is that for the particular type of object we are selecting (radio-loud quasars) we must use the *smaller* of the two possible values in each case. The reason for this is that the source must satisfy *two* criteria to be identified as a radio-loud quasar; it must first be brighter than some radio-frequency flux limit to be a candidate source, *and* subsequent spectroscopy must verify its QSO nature. Failure to take this into account will lead to an underestimate of V/V_{max}.

The one obvious key to making this test work is to be able to determine the value of z_{max} reliably for each source, i.e., the detection limit for a given observation. It is thus worth noting the outcome of making an error in estimating the limiting flux S. If our estimate of S is too low, the associated value of z_{max} will be too large and $\langle V/V_{max} \rangle$ will be systematically too low. Similarly, if S is overestimated, the bias will work the other direction. However, it is also obvious that if we overestimate S for the sample, we will find some sources with $z > z_{max}$, which will tell us right away that we have erred. Thus the tendency will be for us to err in the direction of underestimating S (i.e., believing our detection limit is better than it actually is), and errors are more likely to be in the sense of underestimating rather than overestimating $\langle V/V_{max} \rangle$. This is a useful thing to keep in mind when we consider the results of these tests.

One further complication in use of the V/V_{max} test is that in some cases there is also a *lower* limit to the distance at which a source can be observed. This occurs, for example, in the case of objective prism (or grism) surveys, since a source must have some minimum redshift for the emission lines to accessible. For example, a spectroscopic survey for Lyα emission with a ground-based detector that has a short-wavelength cut-off of 3500 Å will not be able to find any sources with $z \lesssim 1.9$. In such cases, the test is modified to test over only the *accessible* volume V_a (Avni and Bahcall 1980), and the test statistic becomes

$$\frac{V}{V_a} = \frac{V - V_{min}}{V_{max} - V_{min}}, \tag{11.26}$$

where of course V_{min} represents the inner part of the volume which is inaccessible.

While we have emphasized the power of V/V_{max} as a test for a non-uniform density, it is important to remember that it does *not* give us any handle on the functional form of the density as a function of distance, only whether the density is higher or lower locally than farther away (i.e., if $\langle V/V_{max} \rangle > 0.5$, there are more objects at larger distances than at smaller ones). For QSOs, the distances involved are so large that the lookback times are significant, so what we are really testing with the V/V_{max} test is evolution, not spatial homogeneity.

Table 11.1

Selected V/V_{max} Results

Sample	$\langle V/V_{max}\rangle$	N	ref.
3CR QSOs	0.694 ± 0.042	34	1
4C QSOs	0.698 ± 0.024	76	1
PKS QSOs	0.647 ± 0.031	60	1
All above, flat spectrum, $\alpha \leq 0.5$	0.602 ± 0.028	73	1
All above, steep spectrum, $\alpha > 0.5$	0.722 ± 0.018	138	1
X-ray-selected QSOs	0.660 ± 0.062	22	2
X-ray-selected BL Lacs	0.340 ± 0.062	22	3
Grism survey, C IV $(2.0 \lesssim z \lesssim 3.2)$	0.528 ± 0.036	72	4
Grism survey, Lyα $(2.7 \lesssim z \lesssim 4.7)$	0.377 ± 0.028	90	4

References:
1: Wills and Lynds (1978).
2: Weedman (1986).
3: Stocke *et al.* (1991).
4: Schmidt, Schneider, and Gunn (1991).

11.1.4 Results of the Luminosity–Volume Test

Some selected results of the V/V_{max} test are summarized in Table 11.1. Note that the last two entries are for QSOs selected by slitless spectroscopy, and then followed up with high-resolution, higher signal-to-noise ratio slit spectroscopy. The lines selected (Lyα and C IV) are accessible from the ground only within a certain redshift range, so the values of V/V_{max} and the errors given here are adjusted for the absence of low-redshift objects as specified in eq. (11.26).

For both radio-selected (3CR, 4C, and PKS) and X-ray-selected QSOs, $\langle V/V_{max}\rangle$ is significantly greater than 0.5, which strongly indicates that the quasar population evolves with time. Either there were more quasars in the past than there are now (density evolution), or the individual sources are now fainter than they used to be (luminosity evolution). However, it is notable that the spectroscopically selected Lyα QSOs have a value of $\langle V/V_{max}\rangle$ *less* than 0.5, which suggests a change in this trend at higher redshifts.

The results for BL Lac objects whose redshifts have been determined are also given in Table 11.1. In contrast to the QSO population, the space density of BL Lac objects does not appear to increase with redshift, and seems in fact to decline.

11.2 The QSO Luminosity Function

The simple tests discussed in the previous section provide clear evidence for evolution of the population of QSOs per comoving volume over cosmological time scales. This means that the luminosity function must be considered to be a function of z; we simply cannot take a large sample of QSOs regardless of redshift and use them in a luminosity-function calculation. We will need, rather, to consider the distribution of luminosities or absolute magnitudes over relatively restricted redshift intervals. Since we are examining the distribution over two parameters, L and z, a large sample of objects is required to avoid problems of small-number statistics.

Once a large sample of QSOs has been isolated, determination of the luminosity function is in principle fairly straightforward. The absolute magnitude of a QSO at redshift z is given by

$$M = m - A(\ell, b) - K(z) - 5 \log d_{\mathrm{L}} - 25, \tag{11.27}$$

where in the interest of precision we have included explicitly the correction for interstellar extinction $A(\ell, b)$.[†]

If the QSOs are drawn from a 'volume-limited' sample (i.e., all QSOs within some volume $V_{\max}$), the luminosity function is simply computed from

$$\phi(M) \, \Delta M = \sum_{M_i \in (M \pm \Delta M/2)} \frac{1}{V_{\max}} = \frac{N_M}{V_{\max}}, \tag{11.28}$$

where N_M is the number of QSOs in the sample with absolute magnitudes between $M - \Delta M/2$ and $M + \Delta M/2$. More often, QSOs are drawn from 'flux-limited' or 'magnitude-limited' samples; the total volume of space surveyed is thus a function of the absolute magnitude M. We must replace $V_{\max}$ with $V_{\max}(M)$ to account for the fact that more luminous objects can be seen at larger distances than fainter sources and are thus overrepresented in a magnitude-limited sample. Weighting each individual QSO by the reciprocal of the volume over which it could have been found corrects this problem, which is known as the Malmquist bias (cf. pp. 239ff. of Mihalas and Binney 1981). Again, for a correct treatment, we need to normalize by the accessible volume V_a (as in eq. 11.26) for each QSO. Finally, since we have already found evidence that the

---

† In computing absolute magnitudes and luminosities of extragalactic sources, we must always correct for foreground Galactic extinction. The most widely accepted way to determine the foreground extinction is by measuring the H I 21-cm emission column in the direction of the source, at Galactic longitude ℓ and latitude b (e.g., Burstein and Heiles 1982, 1984). From the 21-cm measurement, the foreground reddening can be estimated, basically by assuming a constant neutral hydrogen-to-dust ratio, which at high Galactic latitude corresponds approximately to

$$\frac{N(\mathrm{H\,I})}{E(B - V)} = 4.93 \times 10^{21} \text{ atoms cm}^{-2} \text{ mag}^{-1}$$

(Lockman and Savage 1995). The extinction as a function of wavelength can then be computed from a standard interstellar reddening curve, such as that of Cardelli, Clayton, and Mathis (1989). Of special note are pointed 21-cm observations of brighter AGNs that yield high-accuracy neutral-hydrogen column densities (e.g., Elvis, Lockman, and Wilkes 1989, Lockman and Savage 1995).

space density of QSOs changes with time, we need to compute the luminosity function
as a function of redshift as well, i.e.,

$$\phi(M,z)\,\Delta M = \sum_{\substack{M_i \in (M \pm \Delta M/2) \\ z_i \in (z \pm \Delta z/2)}} \frac{1}{V_a(i)}. \tag{11.29}$$

As suitable samples of AGNs began to be accumulated, it became apparent that at
least at low redshift the QSO luminosity function is steeper at high luminosity than
at low luminosity. This led to a simple parameterization of the luminosity function as
a two-component power law, with different slopes at high and low luminosities and a
'break point' where the slope changes. A similar description has been used in the past
for the luminosity function of normal galaxies (e.g., Abell 1962). A mathematically
superior formulation (since the derivative is not discontinuous at the break point) is
the form used by Marshall (1987)

$$\phi(L,z)\,dL = \phi^* \left\{ \left(\frac{L}{L^*(z)}\right)^{-\alpha} + \left(\frac{L}{L^*(z)}\right)^{-\beta} \right\}^{-1} \frac{dL}{L^*(z)}. \tag{11.30}$$

This equation can also be written equivalently in B magnitudes by defining

$$M_B = M_B^*(z) - 2.5 \log L/L^*(z) \tag{11.31}$$

and using the transformation

$$\phi(M_B,z)\,dM_B = \phi(L,z) \left| \frac{dL}{dM_B} \right| dM_B. \tag{11.32}$$

It is straightforward to show that the equivalent form is

$$\begin{aligned}
\phi(M_B,z)\,dM_B &= \phi_M^* \{ \mathrm{dex}\,[0.4\Delta M_B(z)(\alpha + 1)] + \\
&\quad \mathrm{dex}\,[0.4\Delta M_B(z)(\beta + 1)]\}^{-1} \, dM_B,
\end{aligned} \tag{11.33}$$

where $\phi_M^* = \phi^*(0.4 \ln 10)$. Here $\Delta M_B(z) = M_B - M_B^*(z)$, where both $M_B^*(z)$ and $L^*(z)$
indicate the possibly redshift-dependent break point, or characteristic luminosity. The
redshift dependence of the characteristic luminosity is conventionally modeled as

$$L^*(z) = L_0^*(1+z)^k, \tag{11.34}$$

or equivalently $M_B(z) = M_B^* - 2.5k \log(1+z)$, where L_0^* and M_B^* refer to the characteristic
luminosity extrapolated to zero redshift.

By using a composite catalog of over 700 QSOs, Boyle *et al.* (1991) have obtained
best-fit values for this parameterization for $0.3 \lesssim z \lesssim 2.9$. For the Einstein–de Sitter
case, the best-fit power-law indices are $\alpha = -3.9$ and $\beta = -1.5$ for an assumed
UV/optical SED (which defines the K-correction in eq. 11.27) $F_\nu \propto \nu^{-1/2}$. The
characteristic luminosity is given by

$$M_B(z) = M_B^* - 2.5k \log\left[1 + \sup(z, z_{\max})\right], \tag{11.35}$$

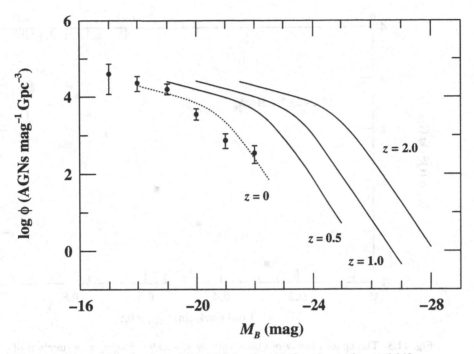

Fig. 11.2. A parameterization of the QSO luminosity function at low redshift $(0.3 \lesssim z \lesssim 2.9)$ by Boyle *et al.* (1991), as in eq. (11.30). This function extrapolated to zero redshift is shown as a dotted line. Also shown as filled circles with appropriate error bars is the Seyfert 1 luminosity function from Cheng *et al.* (1985).

where $M_B^* = -20.9 + 5 \log h_0$ mag and $\sup(z, z_{max})$ refers to the larger of the two quantities z and z_{max}. The characteristic luminosity evolves as $(1 + z)^k$, with $k = 3.45$, up to a maximum redshift $z_{max} = 1.9$, above which it remains constant. In this formulation, the α-term dominates at higher luminosity where the luminosity function slope is $d \log \phi / d \log L \approx \alpha$ and the β-term dominates at low luminosity, where the luminosity function slope is $d \log \phi / d \log L \approx \beta$. The normalization of the luminosity function is given by $\phi_M^* = 5.2 \times 10^3 h_0^3$ QSOs mag^{-1} Gpc^{-1}. We plot this function for several different redshifts in Fig. 11.2.

The luminosity function at $z \gtrsim 3$ is not yet well determined, although most of the surveys of high-redshift QSOs appear to be in good agreement (Kennefick, Djorgovski, and de Carvalho 1995). Current estimates of the space density of the most luminous QSOs ($M_B \geq -25.5 + 5 \log h_0$ mag) as a function of redshift are shown in Fig. 11.3, which shows clearly that the comoving space density declines dramatically (note that the vertical axis is logarithmic) at $z \gtrsim 3$, but there is no sign of an abrupt high-redshift cut-off in the distribution. From this figure, we immediately realize that the era of QSOs occurred at $z \approx 2$; the space density of QSOs before and after is almost minuscule by comparison.

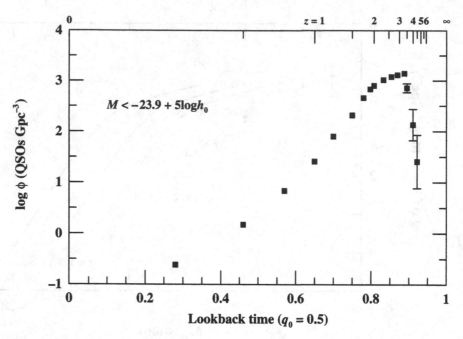

Fig. 11.3. The space density of QSOs with $M < -23.9 + 5\log h_0$ as a function of lookback time (in units of the age of the Universe) for $q_0 = 0.5$ (Warren, Hewett, and Osmer 1994). The redshift scale is also indicated on the top axis. The space density of QSOs drops off dramatically at $z \gtrsim 3$ (i.e., lookback times greater than 0.9). Data courtesy of P. S. Osmer.

11.3   QSO Evolution

While we now have clear evidence that the AGN population evolves with cosmic time, we still have not developed obvious constraints on how individual QSOs change with time. The extreme possibilities are essentially those presented in Chapter 1; either a small fraction of bright galaxies harbor AGNs, and the luminosities of these sources change systematically with time ('luminosity evolution'), or virtually all bright galaxies harbor AGNs, but at any given time most are in 'dormant' states. In the latter case, the fraction of galaxies with AGNs in an 'active' state changes with time ('density evolution'). We will consider here some of the possible consequence of these forms of evolution and what they imply about the individual AGNs.

11.3.1   Mass Accretion on Cosmological Time Scales

In Chapter 3, we computed the mass accretion rates necessary to account for the luminosity of AGNs. The insightful reader may have been disturbed at the implications for the growth of the central masses over the age of the Universe, and we consider this issue further here. We suppose that the luminosity of a QSO is given by its mass accretion rate

$$\dot{M} = \frac{dM}{dt} = \frac{L}{\eta c^2},$$ (11.36)

and that it is radiating at the Eddington limit

$$L = \frac{4\pi G c m_p}{\sigma_e} M.$$ (11.37)

Combining these gives

$$\frac{dM}{dt} = \frac{4\pi G m_p}{c\sigma_e \eta} M,$$ (11.38)

which can be rearranged to

$$\frac{dM}{M} = \frac{4\pi G m_p}{c\sigma_e \eta} dt = \frac{dt}{\tau \eta},$$ (11.39)

where $\tau = 1.43 \times 10^{16}$ sec $= 4.6 \times 10^8$ years. Thus, if η is constant with time, the QSO mass is given by $M = M_0 e^{t/\tau\eta}$, where M_0 is an initial mass which then grows exponentially on a time scale $\tau\eta$. This is often taken to be the upper limit to the rate at which a supermassive black hole can grow, but this is not true if the efficiency is low and the central object is not radiating at the Eddington limit. It is clear already from the luminosity functions shown in the figures that individual QSOs cannot evolve in such a way that they are always radiating at the Eddington limit; if this were true the typical QSO today would be much brighter than the typical QSO at $z = 2$ and this is plainly not the case. Indeed, if QSOs were radiating near the Eddington limit at $z = 2$, they must be radiating at grossly sub-Eddington rates today. This is in fact a strong argument that a thin accretion-disk model, for which $\dot{M}/M_E \lesssim 1$ as noted in Chapter 3, is appropriate for AGNs.

Obviously, we cannot use the simple accretion theory outlined in Chapter 3 to compute the mass accreted by AGNs over the history of the Universe. However, we can compute the total mass accreted by all of the QSOs in a typical comoving volume, without recourse to the details of accretion theory, simply by adding up all of the radiation that has been generated by AGNs over the lifetime of the Universe. And, as shown by Sołtan (1982), this can be done in an elegant fashion that obviates the need to determine the luminosity function as an intermediate step and instead can be related directly to the observed surface density of QSOs.

The quantity we wish to know is the total radiant energy produced per unit comoving volume by QSOs of luminosity L at some cosmic time t, which is

$$E(L, t) \, dL \, dt = L \phi(L, t) \, dL \, dt,$$ (11.40)

where $\phi(L, t)$ is the luminosity function. Rewriting eq. (11.6) to include only sources of a specified flux F (and therefore a specified luminosity L, since z is also specified), the number of sources with flux F in a comoving volume element dV at redshift z is

$$
\begin{aligned}
d\mathcal{N}(F,z)\,dF\,dz &= n(F,z)\,dV(z)\,dL \\
&= \phi(L,z)\frac{dV}{dz}\,dz\,dL,
\end{aligned}
\tag{11.41}
$$

where dV/dz is given by eq. (11.9).

We will also require the connection between cosmic time and redshift, which is given by eq. (9.71),

$$
dt = \frac{da}{H_0 a_0\left[1 - 2q_0 + 2q_0 a_0/a\right]^{1/2}}.
\tag{11.42}
$$

Recalling that $1 + z = a_0/a$, so that $dz = -a_0 da/a^2$, we can write

$$
\frac{da}{a_0} = \frac{a^2 dz}{a_0^2} = \frac{-dz}{(1+z)^2}.
\tag{11.43}
$$

We can combine these two equations to get

$$
dt = \frac{-dz}{H_0(1+z)^2\,(1+2q_0 z)^{1/2}}.
\tag{11.44}
$$

This relationship will allow us to transform the dependence of the luminosity function from time to redshift, i.e.,

$$
\phi(L,t)\,dL\,dt = \phi(L,z)\,dL\,\frac{dt}{dz}\,dz.
\tag{11.45}
$$

We can now proceed with the calculation by dividing eq. (11.41) through by $\delta\Omega$ to obtain the surface density of sources of flux F in dV at z. This yields

$$
\frac{d\mathcal{N}(F,z)}{d\Omega}\,dF\,dz = \phi(L,z)\,dL\,dz\,\frac{1}{\delta\Omega}\frac{dV}{dz}.
\tag{11.46}
$$

Now with $L = 4\pi d_L^2 F$, eq. (11.40) becomes

$$
\begin{aligned}
E(L,t)\,dL\,dt &= L\,\phi(L,t)\,dL\,dt \\
&= L\,\phi(L,z)\,dL\,dz\,\frac{dt}{dz} \\
&= (4\pi d_L^2 F)\left(\frac{dn(F,z)}{d\Omega}dF\,dz\,\frac{\delta\Omega}{dV/dz}\right)\frac{dt}{dz}.
\end{aligned}
\tag{11.47}
$$

Equations (11.44) for dt/dz and (11.9) for dV/dz give us the identity

$$
\begin{aligned}
\frac{d_L^2\,\delta\Omega\,dt}{dV/dz} &= \frac{-d_L^2\,\delta\Omega\,dz}{H_0(1+z)^2\,(1+2q_0 z)^{1/2}}\frac{H_0(1+z)^3\,(1+2q_0 z)^{1/2}}{\delta\Omega\,d_L^2\,c\,dz} \\
&= \frac{-(1+z)\,dz}{c}
\end{aligned}
\tag{11.48}
$$

Thus, the total luminosity emitted by all QSOs from cosmic time $t(z)$ to the present can be obtained by integrating over L and t, i.e.,

$$
\begin{aligned}
E_{\text{rad}} &= \iint E(L, z) \, dL \, dt \\
&= \int \int_{t(z)}^{t_0} L \, \phi(L, t) \, dL \, dt \\
&= \frac{4\pi}{c} \int_0^z (1 + z) dz \int F \frac{d\mathcal{N}(F, z)}{d\Omega} \, dF.
\end{aligned}
\tag{11.49}
$$

The beauty of the above equation is that it does not depend on H_0 or q_0, and depends only on the measured surface density of objects as a function of the measurable quantities F and z. This number can be computed from existing data, with only a bolometric correction needed to account for the broad SEDs of AGNs. Padovani, Burg, and Edelson (1990) find that

$$
E_{\text{rad}} = 1.2 \times 10^{66} \kappa \ \text{ergs} \, \text{Gpc}^{-3} = 3.6 \times 10^{67} \left(\frac{\kappa}{30}\right) \ \text{ergs} \, \text{Gpc}^{-3},
\tag{11.50}
$$

where

$$
\kappa = \frac{\text{Bolometric Luminosity}}{B\text{-band Luminosity}} \approx 30.
\tag{11.51}
$$

The mass accumulated by an AGN over its luminous lifetime is

$$
M_{\text{acc}} = \int \dot{M} dt = \int \frac{L \, dt}{\eta c^2},
\tag{11.52}
$$

so the *cosmological mass density* accumulated by all QSOs in a comoving volume is

$$
\rho_{\text{acc}} = \int \frac{L \, \phi(L, t) \, dt}{\eta c^2} = \frac{E_{\text{rad}}}{\eta c^2} = 2 \times 10^{13} \, \eta^{-1} \ M_{\odot} \, \text{Gpc}^{-3}.
\tag{11.53}
$$

In terms of the critical density (eq. 9.51) $\rho_{\text{crit}} = 2.8 \times 10^{20} h_0^2 \ M_{\odot} \, \text{Gpc}^{-3}$, we see that cosmological mass density in QSOs, $\Omega_{\text{QSOs}} \approx 10^{-7} \eta^{-1} h_0^{-2}$, provides no significant fraction of the closure density. While strictly speaking this applies only to the mass *accreted* by QSOs during their luminous lifetimes, and does not include whatever initial masses they might have, the total mass in the form of QSOs is unlikely to be much higher than this. But the important point is that now we have a handle on the *minimum* cosmological mass density of QSOs, and the required mass per object now depends on how many galaxies the mass is to be distributed over.

11.3.2 Evolution of the AGN Population

How does the luminosity function at $z \approx 2$ evolve to that observed today? The direction of the vector that leads from the $z \approx 2$ luminosity function (Fig. 11.2) to that at $z \approx 0$ is not obvious. As we noted earlier, the two extreme types of evolution that can occur are commonly referred to as density and luminosity evolution. Density evolution can be thought of as a vector in the vertical direction in Fig. 11.2. In other words, the shape of the luminosity function does not change with time; the only change is in the comoving space density of AGNs: there were simply more AGNs in the past than there are now. The absence of very high-luminosity QSOs locally is only because spatially they are so rare. In the case of luminosity evolution, it is supposed that the luminosity function at $z \approx 2$ evolves horizontally in Fig. 11.2. In this scenario, the space density of AGNs is constant with time, but the individual AGNs are all much fainter than they used to be. In this case, the absence of high-luminosity sources locally is because these sources are now much fainter than they were at the epoch corresponding to $z \approx 2$.

Let us consider first the ramifications of pure density evolution. Pure density evolution implies that the probability that an AGN will 'turn off' is independent of luminosity. There are, in this scenario, fewer AGNs now than there were in the past, which means that some of the AGNs that existed in the past have now turned off, and their remnants ('dead quasars') must reside in galaxies that now appear to us as normal.

In Chapter 8, we saw that the evidence is that AGNs reside in bright galaxies. Assuming this is generally true, the maximum number of AGN hosts is thus constrained by the comoving space density of bright galaxies. We take the luminosity function for normal galaxies ϕ_G to be that of Schechter (1976),

$$\phi_G(L)\,dL = \phi_G^* \left(\frac{L}{L^*}\right)^{-5/4} \exp\left(\frac{-L}{L^*}\right) d\left(\frac{L}{L^*}\right), \tag{11.54}$$

where L^* is a fiducial luminosity corresponding to $M_B^* = -19.1 + 5\log h_0$ mag, or equivalently $M_V^* = -20.0 + 5\log h_0$ mag.[†] We take as the normalization constant (Felten 1977)

$$\phi_G^* = 1.76 \times 10^7 h_0^3 \text{ Gpc}^{-3}. \tag{11.55}$$

If we define as 'bright galaxies' those with luminosities $L \geq 0.1L^*$ (i.e., $M_V \leq -17.5 + 5\log h_0$ mag), then the space density of bright galaxies is

$$\begin{aligned}
n_G(L \geq 0.1L^*) &= \int_{0.1L^*}^{\infty} \phi_G(L)\,dL \\
&= \phi_G^* \int_{0.1}^{\infty} \left(\frac{L}{L^*}\right)^{-5/4} \exp\left(\frac{-L}{L^*}\right) d\left(\frac{L}{L^*}\right) \\
&\approx 5.65 \times 10^7 h_0^3 \text{ Gpc}^{-3}.
\end{aligned} \tag{11.56}$$

From Table 7.2, we see that at the current epoch, the total space density of AGNs is about $1.1 \times 10^6 h_0^3$ Gpc^{-3}. The fraction of bright galaxies containing currently active

---

[†] The luminosity function for galaxies is in fact not very well known, although a generalized form of the Schechter function

$$\phi_G(L)\,dL = \phi_G^* \left(\frac{L}{L^*}\right)^{-\alpha} \exp\left(\frac{-L}{L^*}\right) d\left(\frac{L}{L^*}\right)$$

is often adopted. Values of α, L^*, and ϕ_G^* are obtained by fitting this function to the data. More recent estimates of these (interdependent) parameters can be found in Loveday *et al.* (1992) and Marzke, Huchra, and Geller (1994).

AGNs is thus

$$f_{AGN} \approx \frac{1.1 \times 10^6 h_0^3}{5.65 \times 10^7 h_0^3} \approx 0.019. \tag{11.57}$$

Thus, if all bright galaxies contained AGNs, the maximum AGN space density would be increased only by a factor $1/0.019 \approx 50$, or an upperward vertical translation of 1.7 dex in Fig. 11.2. The accreted mass per dormant AGN can be computed by taking the total accumulated mass density (eq. 11.53) and dividing by the space density of host galaxies, i.e.,

$$M_{AGN} = \frac{2 \times 10^{13} \eta^{-1} M_\odot \, \text{Gpc}^{-3}}{5.65 \times 10^7 h_0^3 \, \text{Gpc}^{-3}} \approx 3.5 \times 10^5 \eta^{-1} M_\odot, \tag{11.58}$$

which is not extraordinary, but should be detectable in nearby galaxies (including our own). In fact, evidence is accumulating for the existence of very large central masses (thus far in the range $\sim 2 \times 10^6 M_\odot$ to $\sim 3 \times 10^9 M_\odot$) in local non-active galaxies (reviewed by Kormendy and Richstone 1995).

Examination of Fig. 11.2 shows that a downward translation of the $z = 2$ QSO luminosity function by about 1.7 dex gives a plausible match to the $z = 0$ Seyfert 1 luminosity function over the rather narrow range in luminosity where they overlap. However, such a simple translation significantly over predicts the number of highly luminous QSOs that should be found at low redshift. For example, integrating the total number of QSOs brighter than $M_B = -23.0 + 5 \log h_0$ mag in the $z = 2$ luminosity function, and scaling down by 1.7 dex leads to a prediction that the local space density of such objects should be $\sim 440 h_0^3 \, \text{Gpc}^{-3}$. In the volume extending out to $z = 0.2$ ($V \approx 0.9 h_0^{-3} \, \text{Gpc}^3$), we would expect to find about 400 such objects. In the BQS, which covered about 25% of the entire sky, only 8 such objects were found (Schmidt and Green 1983). On this basis alone, we must consider some form of luminosity evolution, at least at the bright end of the distribution. Of course, examination of the QSO luminosity function at different values of z (as in Fig. 11.2) makes it clear that the break point in the luminosity function varies with z, at least up to $z \approx 1.9$ (eq. 11.35). Clearly then 'pure' density evolution alone will not work – very bright AGNs were simply far too numerous in the past relative to moderately bright AGNs for density evolution alone to be viable.

We therefore proceed to see what would happen in the pure luminosity-evolution scenario. Despite the fact that the density of accreted mass is not large enough to be cosmologically important, it is indeed a significant amount of mass and in the pure luminosity-evolution scenario, all of this mass must be concentrated in existing AGNs. If the current space density of all Seyfert galaxies ($1.1 \times 10^6 h_0^3 \, \text{Gpc}^{-3}$) represents all AGNs that ever existed, the accreted mass per AGN is

$$M_{AGN} = \frac{2 \times 10^{13} \eta^{-1} M_\odot \, \text{Gpc}^{-3}}{1.1 \times 10^6 h_0^3 \, \text{Gpc}^{-3}} \approx 1.8 \times 10^7 h_0^{-3} \eta^{-1} M_\odot, \tag{11.59}$$

which is not inconsistent with the values discussed earlier.

The situation becomes somewhat disturbing, however, if we again consider what happens at the bright end of the AGN luminosity function. Pure luminosity evolution implies that the magnitude of a QSO at $z = 2$ must decrease a factor $(1 + z)^k \approx 47$ (for $k \approx 3.5$) by $z = 0$. As noted earlier, the implication here is that current AGNs are radiating at rates which are grossly sub-Eddington, since they must have accumulated so much mass in the past. For example, consider a QSO at $z = 2$ with $L \approx 10^{48}$ ergs s^{-1}. In a pure luminosity-evolution scenario, the present luminosity of such a source would be

$$L(z = 0) = \frac{L(z = 2)}{47} \approx 2 \times 10^{46} \text{ ergs s}^{-1}, \tag{11.60}$$

where we have assumed the luminosity evolution can be parameterized as in eq. (11.34). The total energy radiated by this QSO over its lifetime (since $z = 2$) is thus

$$\begin{aligned}
E_{\text{rad}} &= \int_{t(z=2)}^{t_0} L(t)\, dt \\
&= \int_{z=0}^{z=2} \frac{L(z)\, dz}{H_0 (1 + z)^2 (1 + 2q_0 z)^{1/2}}, \tag{11.61}
\end{aligned}$$

where we have used eq. (11.44). Considering for simplicity the Einstein–de Sitter case $q_0 = 1/2$ (although the results do not depend strongly on the choice of q_0 over the range of redshifts considered here) and luminosity evolution as in eq. (11.34) with $k \approx 3.5$,

$$\begin{aligned}
E_{\text{rad}} &= \int_{z=0}^{z=2} \frac{L(z = 0)(1 + z)^{3.5}\, dz}{H_0 (1 + z)^{5/2}} \\
&= \frac{L(z = 0)}{H_0} \int_{z=0}^{z=2} (1 + z)\, dz \\
&= \frac{L(z = 0)}{2\, H_0} \left[(1 + z)^2 \big|_{z=0}^{z=2} \right] \\
&= 2.5 \times 10^{64}\, h_0^{-1} \text{ ergs}. \tag{11.62}
\end{aligned}$$

This corresponds to an accreted mass of

$$\begin{aligned}
M_{\text{acc}} &= \int \dot{M}\, dt = \frac{1}{\eta c^2} \int_{t(z=2)}^{t_0} L(t)\, dt = \frac{E_{\text{rad}}}{\eta c^2} \\
&= \frac{2.5 \times 10^{64} h_0^{-1}}{\eta c^2} \text{ g} \\
&= 2.7 \times 10^{43}\, \eta^{-1}\, h_0^{-1} \text{ g} \\
&= 1.4 \times 10^{10}\, \eta^{-1}\, h_0^{-1} \; M_\odot, \tag{11.63}
\end{aligned}$$

which is approximately a galaxy mass! Where are these objects today? The problem is not necessarily alleviated if QSOs are episodic (say, with a 10% duty cycle). This would decrease the mass per QSO, but would necessarily increase the number of galaxies involved.

Both pure density evolution and pure luminosity evolution present serious difficulties, and the real situation must be a hybrid of the two scenarios (e.g., 'luminosity-dependent density evolution'). Pure density evolution distributes the 'dead quasar' mass among a large number of galaxies in a manner that seems to be consistent with the measured central masses of nearby non-active galaxies. However, pure density evolution significantly over-predicts the current space density of high-luminosity QSOs, and thus some luminosity evolution must also be invoked, i.e., the more luminous AGNs are more likely to become dormant than are less luminous AGNs.

12　Quasar Absorption Lines

We have seen in the previous chapters that QSOs are valuable probes of the early Universe because they can be detected at large cosmological distances. The concomitant large lookback times provide a means of studying the Universe during the era when galaxy formation is expected to have occurred. QSOs can also be used in another way as a cosmological probe, namely as background sources against which we see intervening objects. Since the 'sight lines' to individual QSOs are of order Gpc, the chances of finding objects such as galaxies between us and any given QSO are non-negligible. Gas along the line of sight will produce absorption lines in the spectra of QSOs, and the redshift of these absorption lines z_{abs} will reflect the cosmological distance of the absorbing cloud rather than that of the QSO (which will have emission-line redshift z_{em}), so we can expect that QSO spectra will show absorption lines characterized by $z_{abs} < z_{em}$. An 'absorption-line system' consists of a number of absorption lines in a QSO spectrum that are all at very nearly the same redshift z_{abs} and presumably arise in the same absorber. Thus, the objects probed are not the QSOs themselves, but the intervening material that produces the absorption spectrum. We include discussion of absorption-line characteristics (a) because *some* absorption lines actually appear to arise in material associated with the QSOs themselves, (b) because absorption by intervening material modifies the QSO spectrum we observe, and (c) because the study of QSO absorption lines has historically been closely associated with the study of the QSOs themselves.

While the analysis of QSO absorption lines is of enormous potential value in studying the distribution of gas in the Universe, it should become clear that this is very difficult in practice for a number of reasons. First, QSO absorption lines tend to be weak and unresolved, which means that studies of these features involve high spectral resolution, high signal-to-noise ratio observations of apparently faint sources. QSO absorption-line spectroscopy is thus undertaken only on the very largest telescopes, on which observing time is always at a premium. Second, comparative studies of the distribution of a given absorption line of a particular ion as a function of redshift require observing over a large wavelength range, preferably from the rest wavelength of the line λ_0 to the longest wavelength at which it might be detectable, $\lambda_0(1 + z_{em})$. Thus, satellite ultraviolet observations are required to understand the complete distribution in z of the important resonance transitions with rest wavelengths in the UV. The study of QSO absorption lines is an area of tremendous current progress as a result of large amounts of data that are being obtained with *HST* (e.g., Bahcall *et al.* 1993, Bergeron *et al.* 1994) and with the advent of a new generation of large optical telescopes.

12.1 Absorption-Line Physics

The strength of an absorption line centered at wavelength λ_0 can be parameterized by its integrated strength over all wavelengths,

$$\int \sigma(\lambda)\, d\lambda = \frac{\pi e^2}{m_e c} \frac{f \lambda_0^2}{c} = \frac{\lambda_0^4}{8\pi c} \frac{g_2}{g_1} A_{21}, \qquad (12.1)$$

where f is the oscillator strength of the line, and as in Chapter 6, g_1 and g_2 are the statistical weights of the lower and upper levels, respectively, and A_{21} is the Einstein coefficient for the transition. Absorption lines, however, have some finite width on account of both natural broadening, or damping, and Doppler motions of gas particles in the absorbing cloud. Natural broadening occurs because the Heisenberg uncertainty principle and the finite lifetime of the atomic states produce transition energies that are not precisely defined for any single event. Because of broadening, the optical depth through an absorbing cloud is a sensitive function of wavelength near the center of the line. Here we show that even without actually resolving the line profiles, we are able to constrain some of the basic properties of absorbing region from only the equivalent widths, or total strengths, of absorption features (Strömgren 1948).

The relative probability of a photon of wavelength λ being absorbed by an atom with an absorption feature centered at λ_0 is

$$\phi(\lambda)\, d\lambda = \frac{\gamma_i/\pi}{(\lambda - \lambda_0)^2 + \gamma_i^2}\, d\lambda, \qquad (12.2)$$

where γ_i is the damping constant for the line, which in wavelength units† is

$$\gamma_i = \frac{\lambda_0^2}{4\pi c} \sum_{j<i} A_{ij}, \qquad (12.3)$$

where the sum is over all downward radiative transitions (i.e., γ_i is proportional to the reciprocal lifetime of the level i). Equation (12.2) is a normalized ($\int \phi(\lambda)\, d\lambda = 1$) Lorentzian probability distribution that describes the intrinsic absorption-line profile.

In a real astrophysical situation, an ensemble of particles will be characterized by some distribution of line-of-sight velocities that will introduce differential Doppler shifts and thus broaden an absorption line that arises in the gas. For an atom that is moving away from the observer at radial velocity v, the Doppler effect will shift the apparent line center to wavelength $\lambda' = \lambda_0(1 + v/c)$, and thus eq. (12.2) must be

---

† Note that in most compilations, the damping constant is given in unreduced (i.e., excluding the factor of 4π) frequency units, i.e.,

$$\Gamma_i(\text{s}^{-1}) = \sum_{j<i} A_{ij},$$

so as defined here

$$\gamma_i(\text{cm}) = \frac{\lambda_0^2}{4\pi c} \sum_{j<i} A_{ij} = \frac{\lambda_0^2 \Gamma_i}{4\pi c}.$$

modified to

$$\phi(\lambda, v)\, d\lambda = \frac{\gamma_i/\pi}{\left[\lambda - \lambda_0(1 + v/c)\right]^2 + \gamma_i^2}\, d\lambda. \tag{12.4}$$

We consider here the case where the line-of-sight velocity distribution of the ions in the gas is described by a (normalized) Gaussian probability distribution

$$P(v)\, dv = \frac{1}{(2\pi\sigma^2)^{1/2}} e^{-v^2/2\sigma^2}\, dv = \frac{1}{(\pi b^2)^{1/2}} e^{-v^2/b^2}\, dv, \tag{12.5}$$

where b is known as the 'Doppler b parameter'. A Maxwellian (thermal) speed distribution has a Gaussian line-of-sight velocity distribution, and the relationship between the Doppler b parameter and the gas temperature T for purely thermal motions is

$$b = \sqrt{2}\sigma = \left(\frac{2kT}{\mu m_p}\right)^{1/2} = 0.129 \left(\frac{T}{\mu}\right)^{1/2} \text{ km s}^{-1}, \tag{12.6}$$

where μ is the atomic weight of the absorbing ion. By convolving the Lorentz profile, which describes the absorption cross-section per ion, with the velocity distribution of the absorbing ions, we can write the wavelength dependence of the absorption cross-section (per ion) as

$$\sigma(\lambda) = \frac{\pi e^2}{m_e c} \frac{f\lambda_0^2}{c} \frac{1}{(\pi b^2)^{1/2}} \frac{\gamma_i}{\pi} \int_{-\infty}^{\infty} \frac{e^{-v^2/b^2}\, dv}{(\lambda - \lambda_0 - \lambda_0 v/c)^2 + \gamma_i^2}. \tag{12.7}$$

This describes the total absorption profile due to both natural and Doppler broadening and is known as the 'Voigt profile'. Near line center, the Voigt profile is dominated by the Gaussian component, which is referred to as the 'Doppler core'. The Lorentzian component falls off more slowly than the Gaussian with displacement from line center, and results in extended 'damping wings'. This equation can be simplified by substituting $y = v/b$ (so $dv = b\, dy$) and defining a characteristic width of the Doppler core in wavelength units $\Delta\lambda_0 = \lambda_0 b/c$. We can then write $\lambda_0 v/c = \Delta\lambda_0 y$, so that the first term in the denominator of eq. (12.7) becomes $\Delta\lambda_0[(\lambda - \lambda_0)/\Delta\lambda_0) - y]$. We then define the dimensionless terms $a = \gamma_i/\Delta\lambda_0$ and

$$x = \frac{\lambda - \lambda_0}{\Delta\lambda_0}. \tag{12.8}$$

The variable x is the wavelength displacement from line center, in units of the characteristic Doppler width for the distribution. With these substitutions, the absorption cross-section becomes

$$\sigma(\lambda) = \frac{\pi e^2}{m_e c} \frac{f\lambda_0}{b} U(a, x), \tag{12.9}$$

where we have introduced the 'normalized Voigt function'

$$U(a, x) = \frac{a}{\pi^{3/2}} \int_{-\infty}^{\infty} \frac{e^{-y^2}\, dy}{(x - y)^2 + a^2} \tag{12.10}$$

(Hummer 1970).

The equivalent width of an absorption line (see Fig. 12.1) can be defined as

$$W(\lambda_0) = \int \frac{F_c(\lambda) - F_l(\lambda)}{F_c(\lambda)} d\lambda, \tag{12.11}$$

which is identical to eq. (1.9), except for a sign change so that absorption equivalent widths are now positive. Since $F_l(\lambda)/F_c(\lambda) = \exp{-\tau(\lambda)}$, this can be written

$$W(\lambda_0) = \int \left(1 - e^{-\tau(\lambda)}\right) d\lambda, \tag{12.12}$$

where the optical depth is

$$\tau(\lambda) = \int n(\ell)\sigma(\lambda)d\ell = N\sigma(\lambda). \tag{12.13}$$

The integral is over the path length ℓ through an absorbing cloud that has a particle density $n(\ell)$ (cm^{-3}) ions in the lower state. This defines the column density $N = \int n(\ell)\,d\ell$ (cm^{-2}) of absorbing ions through the cloud. Again changing variables as in eq. (12.8), we can write the equivalent width as

$$W(\lambda_0) = \frac{\lambda_0 b}{c} \int \{1 - \exp{[-\tau_0 U(a, x)]}\} dx, \tag{12.14}$$

where the optical depth at line center is

$$\tau_0 = \frac{\pi e^2}{m_e c} \frac{f\lambda_0}{b} N. \tag{12.15}$$

Equation (12.14) defines the relationship between τ_0 (which in turn depends on b and N) and the observed equivalent width $W(\lambda_0)$ that is known as the 'curve of growth', since it describes how $W(\lambda_0)$ 'grows' with increasing τ_0. The basic characteristics of the curve of growth can be seen by approximating the behavior of eq. (12.14) for various regimes of τ_0:

Unsaturated line: A line is said to be 'unsaturated' if the residual (unabsorbed) intensity at line center is non-negligible. In the case $\tau_0 \ll 1$, $\tau(\lambda)$ is small everywhere, and a first-order expansion of eq. (12.14) yields $W(\lambda_0) \propto \tau_0$. This is referred to as the 'linear' part of the curve of growth.

Saturated line: A line is said to be saturated when the transmitted intensity at line center is virtually zero. Essentially no photons at line center make it through the absorbing cloud. However, at slight displacements from line center, in the Doppler wings of the line, the transmission may still be non-zero. Since the probability of a photon being absorbed drops off exponentially through the Doppler core of the line, for a simple analysis we can suppose that the line transmission is effectively zero for some $x < x^*$ and effectively 100% for $x > x^*$ (i.e., we approximate the Doppler core of the line as a rectangular function of half width x^*). The value of x^* is defined by

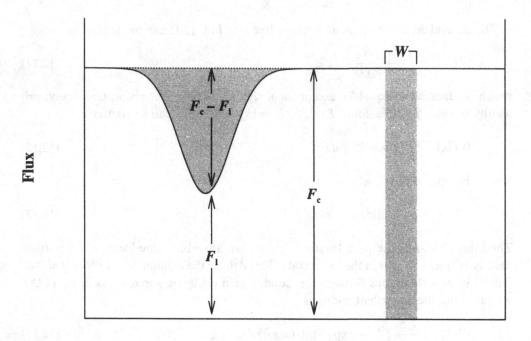

Wavelength

Fig. 12.1. The equivalent width W of an absorption line, as defined by eq. (12.11). The shaded areas of the two lines shown are the same; the amount of energy absorbed by the Gaussian profile on the left is equivalent to that absorbed by an idealized rectangular absorption feature of width W, which is measured in wavelength units.

the condition

$$\tau_0 U(a, x^*) = 1. \tag{12.16}$$

The Doppler core can be thought of as the regime where $x \approx y \gg a$ in eq. (12.10). If we approximate the exponential as having constant value $\exp -x^2$ in this regime, then $U(a, x^*) \approx (\exp -x^{*2})/\sqrt{\pi}$, and using condition eq. (12.16), we see that $W \propto (\ln \tau_0)^{1/2}$. Thus, in the regime where the center of the line but not the entire Doppler core is saturated, the line equivalent width is a very insensitive function of τ_0; thus the 'Doppler part' of the curve of growth is also sometimes referred to as the 'flat part' of the curve because $W(\lambda_0)$ grows only very slowly with τ_0.

Heavily saturated line: Once the optical depth at line center becomes very large, the entire Doppler core of the line is saturated. The only unsaturated part of the line is far out in the low-opacity damping wings of the line. Far out in the wings, $(x - y)^2 \approx x^2 \gg a^2$, and eq. (12.10) becomes approximately $U(a, x^*) \approx (a/\pi x^{*2})$. Again using condition eq. (12.16), we have $W(\lambda_0) \propto \tau_0^{1/2}$. Thus, the 'damping part' of the curve of growth is also sometimes known as the 'square-root part' of the curve.

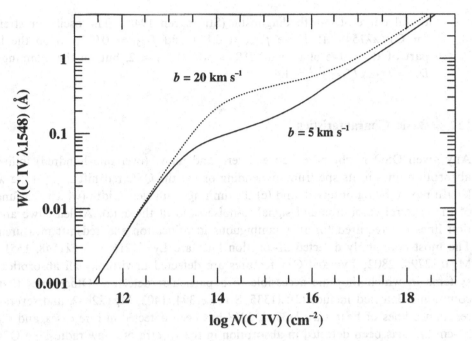

Fig. 12.2. Representative curve of growth for C IV λ1548, for two different values of the Doppler b parameter. The curve for the lower value of b saturates first (eq. 12.15) because for a fixed value of N this has the larger number of ions per unit dv.

In Fig. 12.2, we show a sample curve of growth for the C IV λ1548 line for low and moderate values of b. It is clear from this diagram that measurement of the equivalent width of a single line gives very limited information because there are generally a family of solutions (N, b) that yield any value of $W(\lambda_0)$. On the linear part of the curve of growth, the relationship between $W(\lambda_0)$ and τ_0 is completely degenerate since $W(\lambda_0)$ depends linearly on the ratio N/b and N and b remain separately unconstrained. In order to use the curve of growth to determine or set limits on N and b, the equivalent widths of several different lines, preferably from the same ion and on different parts of the curve of growth, need to be measured. In the absence of information on several different lines, it is still sometimes possible to obtain constraints on N and b:

- Line profiles can be fitted with Voigt profiles to determine b values, and then b and $W(\lambda_0)$ together yield N. This can be done at even moderate resolution (say, $R \approx 5000$) if the lines are on the damping part of the curve of growth, but otherwise high resolution ($R \gtrsim 10\,000$–$20\,000$) is required (e.g., Rauch *et al.* 1992).

- Some of the strongest lines in QSO absorption spectra are resonance doublets (e.g., C IV $\lambda\lambda$1548, 1551, and Mg II λ2795, 2802), which have different oscillator strengths and thus saturate at slightly different column densities. Thus the ratio of their equivalent widths, known as the 'doublet ratio' *DR*, provides a useful curve-of-growth diagnostic (Strömgren 1948). The oscillator strengths for C IV $\lambda\lambda$1548, 1551 are $f_{1548} = 0.194$ and $f_{1551} = 0.097$, so on the linear part of the curve of growth, $DR \approx f_{1548}/f_{1551} \approx 2$, but on the damping part $DR = (f_{1548}/f_{1551})^{1/2} \approx 1.4$.

12.2   Basic Characteristics

Any given QSO might have between zero and many (over one hundred) individual absorption lines in its spectrum, depending on (a) the QSO redshift z_{em}, (b) the wavelength region being observed, and (c) the limiting equivalent width (which is a function of the spectral resolution and signal-to-noise ratio of the data). At least two absorption lines are required for an unambiguous identification and redshift measurement. The most commonly detected absorption lines are Lyα λ1216, C IV $\lambda\lambda$1548, 1551, and Mg II λ2795, 2802; Lyα and C IV features are detected in virtually all absorption-line systems in which they are accessible to a particular detector. Other lines that are commonly detected include C II λ1335, Si IV $\lambda\lambda$1394, 1403, Mg I λ2852, and several UV resonance lines of Fe II. Ca II $\lambda\lambda$3933, 3968 has been detected in rare cases, and the H I 21-cm line has been detected in absorption in the spectra of a few radio-loud QSOs.

Most QSO absorption systems fall into one of the following categories (Weymann, Carswell, and Smith 1981, Sargent 1988):

Heavy-element (or 'metal-line') systems: These absorption systems consist of lines of ionized and neutral metals, as well as Lyα when it is accessible. In general, these lines are narrow and unresolved except at the very highest spectral resolution. Curve-of-growth analysis suggests widths usually of order tens of km s^{-1}, which is too large to attribute to purely thermal motions within the clouds, and thus suggests that individual absorption systems arise in several clouds that are close in velocity space. The widths presumably reflect the bulk motions of these clouds. Additional support for this scenario is provided (a) by structure on scales of a few km s^{-1} seen in some 21-cm absorption lines and (b) by close splittings (of order hundreds of km s^{-1}) seen in some optical absorption systems. The equivalent widths of most detected lines are of order a few Å down to the weakest detection limits of several tens of mÅ. The inferred column densities are typically in the range $\sim 10^{17}$–10^{21} cm^{-2}. There are some claims that metals are slightly underabundant relative to solar values. An important subclass of the heavy-element systems are 'damped Lyα systems', which are systems with such large column densities that the damping wings of Lyα are optically thick, so that the Lyα absorption is strong and broad (e.g., Wolfe *et al.* 1986). The Lyα equivalent widths of these systems are typically $W(\text{Ly}\alpha) \gtrsim 10$ Å. These systems are thought to arise on sight lines which pass through galactic disks. A metal-line system with a damped Lyα line is shown in Fig. 12.3. Other closely related absorption-line systems are those in which the Lyman continuum is detected in absorption, the 'Lyman-limit systems' (Tytler 1982), as shown in Fig. 12.4. These systems are clearly due to intervening gas

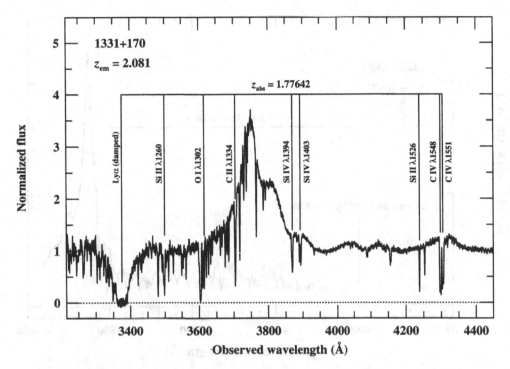

Fig. 12.3. Many of the strongest absorption lines in 1331+170 ($z_{em} = 2.081$) are identifiable in a single redshift system at $z_{abs} = 1.77642$ (Chaffee, Black, and Foltz 1988). The strong emission line at $\sim 3750\,\text{Å}$ is Lyα. The column density in low-ionization species is high enough to produce strong damping wings on Lyα. The spectrum has been normalized to a value of unity for the locally unattenuated continuum flux. Data courtesy of C. B. Foltz.

that is unassociated with the QSOs since they usually appear at redshifts significantly less than z_{em}. Thus, the observed Lyman discontinuities do not arise in either the AGN accretion disk (§4.1) or in the BLR (§5.2). Molecular absorption in metal-line systems is apparently very rare, although H_2 absorption has been found in at least one damped Lyα system (Levshakov and Varshalovich 1985, Foltz, Chaffee, and Black 1988). There is, however, substantial evidence for dust in such systems, both through reddening of the continuum (Pei, Fall, and Bechtold 1991) and through abundances analyses (Pettini *et al.* 1994) that indicate depletion of gas-phase chromium relative to zinc, presumably because the former has been removed from the gas to form dust. These abundance analyses also yield metallicity estimates of $\sim 0.1 Z_\odot$ for the damped Lyα clouds. The presence of substantial amounts of dust in such clouds might have a potentially serious effect on our ability to detect high-redshift QSOs that lie behind them (Fall and Pei 1993).

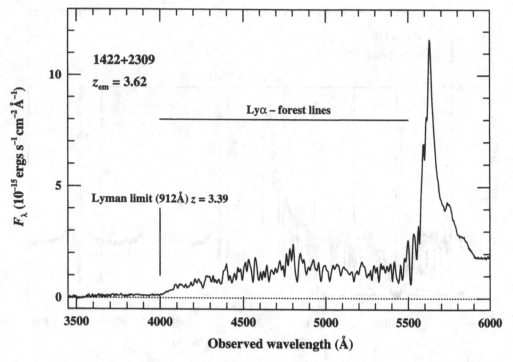

Fig. 12.4. The spectrum of the QSO 1422+2309 cuts off dramatically at wavelengths shorter than 4000 Å because of an absorption system at $z = 3.39$ that is optically thick at rest wavelengths shortward of 912 Å (Patnaik *et al.* 1992). This is an example of a 'Lyman limit' absorption system. Data courtesy of C. B. Foltz.

Lyα-forest systems: At wavelengths shortward of the Lyα emission line in QSOs, the density of absorption systems (i.e., number of absorption lines per unit wavelength) increases dramatically, as first noticed by Lynds (1971) and surveyed in detail by Sargent *et al.* (1980). Nearly all of these lines are attributed to Lyα absorption, as these lines often have no counterparts at wavelengths longer than the Lyα emission line. In high-redshift QSOs, the continuum shortward of Lyα emission is completely riddled with absorption features (see Figs. 12.4 and 12.5), and these lines are collectively referred to as the 'Lyα forest' (Weymann, Carswell, and Smith 1981) on account of the high density per unit wavelength of such features. Some of these systems may well be metal-line systems in which the column densities are just too low to produce detectable absorption in any line except Lyα. However, the inferred neutral-H column densities for *some* of these systems are as high as $\sim 10^{17}$ cm^{-2}, and the absence of metal lines in some of the higher column-density systems implies that metals are underabundant relative to solar values, i.e., typically $Z \lesssim 0.01 Z_\odot$ (e.g., Tytler and Fan 1994). Low-redshift counterparts to Lyα forest lines have been detected with *HST* (Bahcall *et al.* 1991, Morris *et al.* 1991).

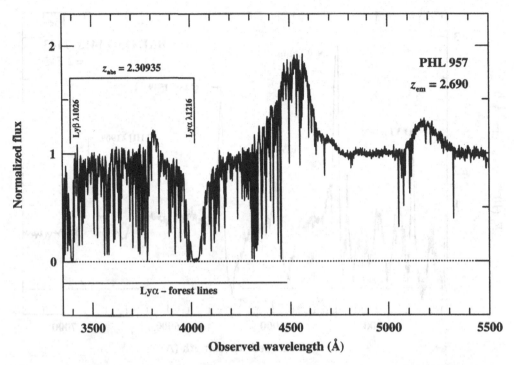

Fig. 12.5. The QSO PHL 957 has a rich absorption spectrum, including a damped Lyα line at $z = 2.309\,35$, in which Lyβ is also strong (Black, Chaffee, and Foltz 1987). The prominent emission lines in this spectrum are Lyα near 4550 Å and Si IV near 5200 Å. Most of the many lines shortward of the Lyα emission line are Lyα absorption lines, the 'Lyα forest' absorption systems. The spectrum has been normalized to a value of unity for the locally unattenuated continuum flux. Data courtesy of C. B. Foltz.

Broad Absorption Lines (BALs): Broad ($\sim 10^4\,\mathrm{km\,s^{-1}}$) absorption features are detected in the shortward wings of resonance lines of some QSOs (Weymann, Carswell, and Smith 1981). These absorption features are of the P Cygni type, although in some cases the absorption features are 'detached' from the emission line by as much as $\sim 30\,000\,\mathrm{km\,s^{-1}}$. The absorption is always at wavelengths shortward of line center, which indicates that the absorbing gas is flowing outward from the nucleus, and the high ionization level and high outflow velocities of the gas strongly suggest that these systems are closely associated with the nuclear regions. The broad absorption features are sometimes found to be variable (e.g., Barlow *et al.* 1992). A sample BAL QSO is shown in Fig. 12.6.

Some absorption lines are detected at redshifts slightly larger than z_{em} (Weymann *et al.* 1977). The inferred relative velocities of such systems are typically $\Delta v \lesssim 3000\,\mathrm{km\,s^{-1}}$. The $z_{\mathrm{abs}} > z_{\mathrm{em}}$ phenomenon is thought to be attributable to some combination of QSO and absorber peculiar velocities relative to the Hubble expansion and intrinsic

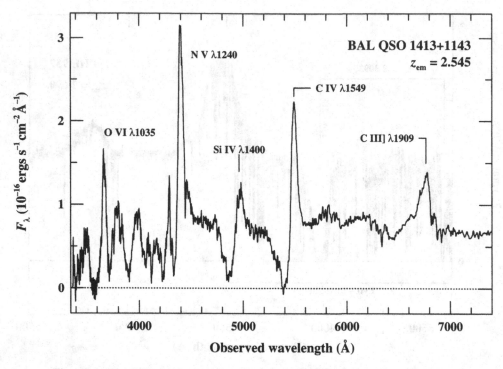

Fig. 12.6. The QSO 1413+1143 is an example of a broad absorption-line (BAL) QSO (Weymann *et al.* 1991). The resonance lines of abundant ions are strongly absorbed on the shortward side of the line, which indicates that the absorbing gas is flowing outward from the active nucleus. Lyα absorption is so strong that Lyα emission is not detectable. Data courtesy of C. B. Foltz.

wavelength shifts of the broad emission lines relative to their systemic redshifts (Gaskell 1982; §5.4.2).

12.3   Broad Absorption-Line QSOs

While broad absorption lines in QSOs are relatively rare, they are important because they provide yet another probe of the central regions of AGNs. The first known BAL QSO was PHL 5200 (Lynds 1967), and very few additional examples were discovered prior to the widespread use of objective-prism spectroscopy as a means of isolating QSOs. An atlas of 72 BAL spectra has been published by Korista *et al.* (1993).

BAL systems are detected in approximately 9% of all QSOs with $z \gtrsim 1.5$ (Weymann *et al.* 1991). The presence of these strong absorption troughs significantly depresses the observed SED, which makes detection of BAL QSOs less likely than otherwise similar but unabsorbed QSOs. Correction for this bias suggests that the actual incidence of the BAL phenomenon is about 12% (Foltz *et al.* 1990). BALs are found only in the spectra of *radio-quiet* QSOs, never in the spectra of strong radio sources (Stocke *et al.* 1992). Most BAL systems are dominated by high-ionization species, especially C IV $\lambda\lambda$1548, 1551, Si IV $\lambda\lambda$1394, 1403, and N V $\lambda\lambda$1239, 1243. About 15% of BAL systems are low ionization-level systems in which Mg II $\lambda\lambda$2796, 2803, Al II λ1671, Al III $\lambda\lambda$1855, 1863, and C II λ1335 are also prominent (Voit, Weymann, and Korista 1993). Evidence for dust in at least one low-ionization BAL system has been reported (Sprayberry and Foltz 1992).

It is not known with certainty whether BAL QSOs are intrinsically very rare, or if alternatively the BAL region occurs in all QSOs, but covers only a small fraction of the sky as seen from the central source. At least two tests have been undertaken in an attempt to address this question. The first test is a statistical comparison of the emission-line spectra of BAL and non-BAL QSOs to see if the two types are drawn from the same parent distribution. Results of this test have been mixed. When QSOs drawn from a single large survey are compared, the emission-line properties of the two different types of QSOs appear to be identical (Weymann *et al.* 1991), which argues that there is no intrinsic difference between the types. However, some subsets (e.g., QSOs with low ionization-level BAL systems) *may* show some differences, so the results are currently somewhat ambiguous. A second test is to examine the ratio of total emission equivalent width to total absorption equivalent width $W_{em}(\lambda_0)/W_{abs}(\lambda_0)$ for a given line. If the entire line-emitting region is surrounded by BAL clouds that simply scatter line photons (so that the total number of line photons is conserved), then this ratio must be unity. In other words, if photons are conserved, we expect as many photons to be scattered out of the line of sight as are scattered into it. If $W_{em}(\lambda_0)/W_{abs}(\lambda_0) < 1$, then we conclude that there is more scattering of photons out of the line of sight and therefore that we lie along a preferred direction, i.e., the covering factor is low. In general, $W_{em}(\lambda_0)/W_{abs}(\lambda_0)$ is indeed less than unity, which suggests that the covering factor for BALs is low (e.g., Hamann, Korista, and Morris 1993). Again, the results of this test are somewhat ambiguous because interpretation depends on the model assumptions (such as no destruction of line photons).

The possible relationship of the BAL gas to the BLR is unclear. Strong N V λ1240 absorption in BAL QSOs often nearly obliterates the Lyα emission line, which argues that the BAL gas lies outside the BLR and covers nearly all of the line-emitting region as seen by the observer. Estimation of the mass outflow rates from BALs is highly model dependent, but values of several solar masses per year seem to be indicated (Turnshek 1988), if the abundances are assumed to be solar.

While broad absorption features are not detected in low-luminosity AGNs, many Seyfert galaxies are known to have *narrow* and generally weak absorption features on the shortward side of some emission lines (e.g., Fig. 1.2). Also, while BALs are not seen in radio-loud objects, there have been some indications that narrow C IV absorption features occur close to the emission-line redshift in radio-loud QSOs with a rate of incidence higher than expected if these systems arise in unrelated foreground objects (e.g., Foltz *et al.* 1986). Whether or not all of these phenomena are related is not known.

12.4   Absorption-Line Statistics

12.4.1 The Comoving Number Density of Absorbers

The number of absorption lines we expect along a line of sight to a QSO is given by the usual formula, which is the product of the absorption cross-section, the number density of absorbers, and the path length. The number of absorbers encountered per unit path length at redshift z is thus

$$dN(z) = n(z)\,\sigma(z)\,d\ell(z), \tag{12.17}$$

where $d\ell(z)$ is in this case an element of *proper length*, which from eqs. (9.19) and (9.6) is

$$d\ell(z) = \frac{a(t)dr}{(1 - kr^2/R^2)^{1/2}} = c\,dt. \tag{12.18}$$

The number of absorption lines per unit redshift along a sight line to a QSO is thus

$$\begin{aligned} dN(z) &= n(z)\,\sigma(z)\,c\,dt \\ &= n(z)\,\sigma(z)\,c\,\frac{dt}{dz}\,dz, \end{aligned} \tag{12.19}$$

where $n(z)$ is the number density of absorbers at z and $\sigma(z)$ is their cross-section for producing absorption lines.† We will assume that the cross-section of the absorbers has constant value σ_0, and that the comoving space density is also constant, so that $n(z) = n_0(1 + z)^3$. With eq. (11.44) we thus obtain

$$\begin{aligned} \frac{dN(z)}{dz} &= \left[n_0(1 + z)^3\right]\sigma_0\left[\frac{c}{H_0\,(1 + z)^2\,(1 + 2q_0 z)^{1/2}}\right] \\ &= \frac{c\,\sigma_0\,n_0}{H_0}\,\frac{(1 + z)}{(1 + 2q_0 z)^{1/2}}. \end{aligned} \tag{12.20}$$

While we have considered σ_0 to be constant and n_0 to be constant per unit comoving volume, it is straightforward to consider evolution in either of these parameters by letting them be functions of z. Indeed, it would be rather extraordinary to find no evolution in either of these parameters as we probe back to epochs corresponding to a small fraction of the Hubble time. In studies of the z-dependence of QSO absorption lines, it is customary to parameterize the evolution of the number density of systems as

$$\frac{dN}{dz} = \left(\frac{dN}{dz}\right)_0 (1 + z)^\gamma. \tag{12.21}$$

From eq. (12.20), we see that in the case of *no evolution*, $dN/dz \propto (1 + z)$ for the case $q_0 = 0$ and $dN/dz \propto (1 + z)^{1/2}$ for the case $q_0 = 1/2$. Thus, values of $\gamma > 1$ would imply real evolution in the sense that either the comoving density or cross-section of absorbers is larger at higher z. It must be emphasized that there is no good reason

---

† Carefully note that this is the cross-section per absorbing cloud, rather than the cross-section per ion that was discussed in the curve-of-growth analysis.

to assume a power-law form for the evolution. The density of absorption systems as a function of z is only strictly a power law for no evolution in the special cases $q_0 = 0$ and $q_0 = 1/2$; eq. (12.21) provides a simple parametric form for analyzing the absorption-line statistics. The constant $(dN/dz)_0$, which is a commonly quoted QSO absorption-line statistic, is the number of absorption systems per unit redshift, normalized to $z = 0$. By comparison with eq. (12.20), we see that

$$\left(\frac{dN}{dz}\right)_0 = \frac{c\,\sigma_0\,n_0}{H_0}. \tag{12.22}$$

The normalization of eq. (12.21) is sometimes expressed in terms of $\ell_0 = (\sigma_0 n_0)^{-1}$, the mean free path between absorbers at the current epoch, i.e.,

$$\ell_0 = \frac{c}{H_0}\left(\frac{dN}{dz}\right)_0^{-1}. \tag{12.23}$$

For a sample of N_{QSOs}, the total number of absorption systems we expect to detect is

$$N_{\text{abs}} = \sum_{i=1}^{N_{\text{QSOs}}} \int_{z_{\min}(i)}^{z_{\max}(i)} \frac{dN_i}{dz}\, dz, \tag{12.24}$$

where $z_{\min}(i)$ and $z_{\max}(i)$ are respectively the minimum and maximum redshift at which the particular absorption feature could in principle have been detected in the ith QSO. In practice, the parameter that we extract from the observational data is $dN(z)/dz$. In some redshift interval Δz centered at $\langle z \rangle$, the average number of absorption systems detected in the QSOs surveyed is $\overline{N(\langle z \rangle)}$, and we approximate

$$\frac{dN(z)}{dz} \approx \frac{\overline{N(\langle z \rangle)}}{\Delta z}. \tag{12.25}$$

By fitting eq. (12.21) to these data, estimates of $(dN/dz)_0$ and γ are obtained. Results for some of the common types of absorption systems are given in Table 12.1.

The values of γ for the frequently occurring lines reveal that convincing evidence for evolution of absorbers is seen only in the *strongest* Lyα forest and Mg II systems. The weaker systems, which are numerically dominant, show no strong evidence for evolution in comoving density or cross-section. The situation with C IV is complicated; the density of absorbers apparently *decreases* with z. This has been attributed to lower metallicities at high z (Sargent, Boksenberg, and Steidel 1988).

It is important that the absorption systems included in any such statistical analysis are selected in a consistent way in order to provide a homogeneous sample. For this reason, line selection is usually carried out with objective automated search algorithms (e.g., Schneider *et al.* 1993). As a simple and obvious example of a selection effect that must be allowed for, an absorber of given column density, b value, chemical abundance, and ionization level will produce a given absorption line of fixed *rest-frame* equivalent width W_0. The *observed* equivalent width, however, will be

$$W_{\text{obs}} = W_0(1 + z_{\text{abs}}) \tag{12.26}$$

Table 12.1
QSO Absorption-Line Statistics

Absorption Feature	$(dN/dz)_0$	γ	R/R_H	ref.
Lyα Forest				
($z > 1.6, W_0 > 0.32$ Å)	3.93	1.89 ± 0.28	7.3	1
($z > 1.6, W_0 > 0.16$ Å)	19.2	1.32 ± 0.24	16	1
Damped Lyα λ1216	0.059	1.15 ± 0.55	0.89	2
($0 < z < 3.5$)				
Lyman Limit	0.25	1.50 ± 0.39	1.8	3
($0.3 < z < 4.1$)				
C IV $\lambda\lambda$1548, 1551				
($z \approx 2, W_0 > 0.30$ Å)	1.31	-1.84 ± 0.68	4.2	4
($z \approx 2, W_0 > 0.15$ Å)	2.44	-1.26 ± 0.56	5.8	4
Mg II $\lambda\lambda$2798, 2802				
($z \approx 1, W_0 > 1.00$ Å)	1.19	2.47 ± 0.68	4.0	5
($z \approx 1, W_0 > 0.60$ Å)	1.36	1.11 ± 0.46	4.3	5

References:
1: Bechtold (1994).
2: Lanzetta, Wolfe, and Turnshek (1995).
3: Stengler-Larrea *et al.* (1995).
4: Steidel (1990).
5: Aldcroft, Bechtold, and Elvis (1994).

(Table 9.1). Thus, the detection limit for absorption features decreases with z; absorption features of a given strength are easier to detect at larger z, and this must be taken into account in computing absorption-line statistics. This is usually done by considering only absorption features that are strong enough to have been detected at any redshift surveyed. Similar selection effects, such as the possibility of blended absorption features, must also be considered in a statistical analysis.

It is also important in computing absorption-line statistics for the purpose of studying intervening objects to avoid regions of the QSO spectrum which is populated by absorbers that are physically close to the QSO. The number of absorbers per unit redshift interval can change for $z_{abs} \approx z_{em}$, since there might be a statistical excess of sources in this range because there are additional clouds that are somehow associated with either the QSOs themselves or the QSO environment (e.g., Foltz *et al.* 1986). Conversely, as is observed in the case of Lyα forest lines, the number of lines may *decrease* at $z_{abs} \approx z_{em}$, on account of the increased ionization level of clouds near the QSO (Weymann, Carswell, and Smith 1981, Murdoch *et al.* 1986); this is variously known as the 'proximity effect' or 'inverse effect' (because the otherwise increasing trend in dN/dz reverses). The effect is usually removed from the data by rejecting absorption systems that are within some fixed velocity relative to the QSO itself. As seen from Earth, but in the rest frame of the QSO, a gas cloud moving towards the observer at velocity $v \equiv \beta c$ produces an absorption line at a wavelength

$$\lambda' = \frac{(1-\beta)}{(1-\beta^2)^{1/2}} \lambda_0, \tag{12.27}$$

where λ_0 is the rest wavelength of the line. In the observer's reference frame, the absorption feature occurs at

$$\lambda_{abs} = (1 + z_{em})\lambda' = (1 + z_{em})\frac{(1-\beta)}{(1-\beta^2)^{1/2}} \lambda_0, \tag{12.28}$$

and noting that by definition $\lambda_{abs}/\lambda_0 = 1 + z_{abs}$, it is easy to show that

$$\beta = \frac{(1 + z_{em})^2 - (1 + z_{abs})^2}{(1 + z_{em})^2 + (1 + z_{abs})^2}. \tag{12.29}$$

Possible biases due to absorption systems arising close to a QSO are usually removed by adopting a value of z_{max} that is typically at $\beta \approx 0.1$ less than z_{em}.

12.4.2 Characteristics of the Absorbers

The distribution of rest-frame equivalent widths W as a function of redshift for Lyα-forest lines can be described by an exponential distribution of the form

$$n(W) \equiv \frac{\partial^2 N}{\partial z \, \partial W} = \frac{N^*}{W^*} \exp\left(-W/W^*\right) \tag{12.30}$$

(Sargent *et al.* 1980, Murdoch *et al.* 1986). The characteristic equivalent width is typically found to be $W^* \approx 0.3$ Å.

As we have seen, W has a complicated dependence on the column density N and Doppler parameter b, which are the physically more interesting parameters. The distribution of b values for Lyα-forest lines (Carswell 1989) can be plausibly described by a truncated Gaussian distribution

$$P(b)\,db = \exp\left[\frac{-(b-b_0)^2}{2b_*^2}\right] db, \tag{12.31}$$

for $b > 0$, with fitted values $b_0 \approx 32\,\mathrm{km\,s^{-1}}$ and $b_*^2 \approx 23\,\mathrm{km\,s^{-1}}$, which yields a mean value $\langle b \rangle \approx 36\,\mathrm{km\,s^{-1}}$ (Press and Rybicki 1993).

The distribution of neutral hydrogen column densities in Lyα forest clouds per unit

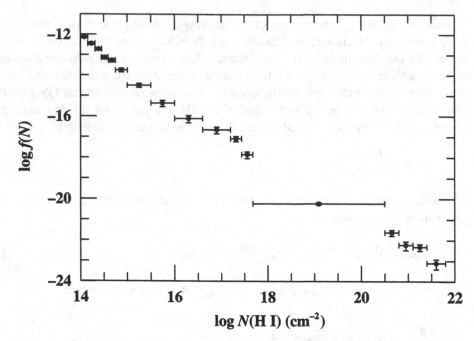

Fig. 12.7. The distribution of QSO absorption-line systems as a function of inferred H I column density, as expressed in eq. (12.32), from Petitjean *et al.* (1993). A single power law of index $\beta \approx 1.5$ is very crude approximation to this function. Data courtesy of D. H. Weinberg.

redshift per unit $N(\mathrm{H\,I})$ is very crudely described by a power-law distribution

$$f(N) \equiv \frac{\partial^2 N}{\partial z \, \partial N(\mathrm{H\,I})} \approx B\, N^{-\beta}, \tag{12.32}$$

where B and β are empirically determined constants. Tabulated values of this function are shown in Fig. 12.7, based on a compilation by Petitjean *et al.* (1993). Over several orders of magnitude in $N(\mathrm{H\,I})$, $\beta \approx 1.5$ is a reasonable description of $f(N)$, although such a simple function is formally a poor fit primarily on account of structure in $f(N)$ in the neighborhood of $N(\mathrm{H\,I}) \approx 10^{15}\,\mathrm{cm}^{-2}$.

12.5   Galaxies as Absorbers

Suppose that the QSO absorption lines arise in material directly associated with galaxies, such as dark halos. In this case, we know *a priori* the number of absorbers from the galaxy luminosity function and we can thus compute the cross-section per galaxy required to account for the observed incidence of absorption lines. The number of galaxies at the present epoch is given by the Schechter luminosity function (eq. 11.54). For lack of any other information, we assume that the absorption cross-sections of galaxies will scale as their optical sizes, so we can use the Holmberg (1975) relationship for the radii of galaxies, which we write in the form

$$R(L) = R^* \left(\frac{L}{L^*} \right)^{5/12}.$$

(12.33)

Here R^* is the radius of a galaxy of fiducial luminosity L^*; the corresponding *optical* radius for such a galaxy, measured at the 26.5-mag arcsec^{-2} isophote in B (i.e., at about 1% the surface brightness of the night sky at its darkest), is called the Holmberg radius R_H and is about $17h_0^{-1}$ kpc for a galaxy of luminosity L^*. The cross-section of a galaxy is assumed to be $\pi R^2 \propto L^{5/6}$. We thus compute the mean free path as

$$\begin{aligned}
\ell_0^{-1} &= \int_0^\infty \sigma(L)\, \phi_G(L)\, dL \\
&= \phi_G^* \pi R^{*2} \int_0^\infty \left(\frac{L}{L^*} \right)^{5/6} \left(\frac{L}{L^*} \right)^{-5/4} \exp\left(\frac{-L}{L^*} \right) d(L/L^*) \\
&= \phi_G^* \pi R^{*2} \Gamma(7/12) \\
&= 4.81 \, \phi_G^* R^{*2},
\end{aligned}$$

(12.34)

where $\Gamma(x)$ is the gamma function.† Solving this with the numerical values given above and in eq. (11.55) yields

$$\ell_0 \approx \frac{41 \, h_0^{-1}}{(R^*/R_H(L^*))^2} \; \text{Gpc}.$$

(12.35)

As shown in the previous section, the observable density of absorption lines can be expressed in terms of the mean free path ℓ_0, which ranges from $\sim 160h_0^{-1}$ Mpc for the Lyα-forest systems with $W_0(\text{Ly}\alpha) \geq 0.16$ Å to $\sim 52h_0^{-1}$ Gpc for the damped Lyα systems. We can use these observed values to solve eq. (12.35) for the typical radius of the gaseous disks of galaxies, relative to their optical sizes, assuming that indeed it is such disks that produce the observed absorption lines. These values are also given in Table 12.1, where it is seen that the metal-rich gaseous halos of galaxies must exceed their optical sizes by a factor of a few at least to account for all of the absorption features detected in QSOs; only the damped Lyα systems are sufficiently few in number that they can be accounted for plausibly by known galaxies. The details of this calculation can be changed, for example, by including only spiral galaxies and by assuming the gas distribution is flattened rather than spherical, and this tends to exacerbate the discrepancy between the observed optical sizes and the inferred absorption cross-sections of galaxies. The implication of this calculation, that heavy elements must exist at very large galactocentric radii, presents a challenge to our understanding of

---

† The gamma function is defined as

$$\Gamma(x) = \int_0^\infty t^{x-1} e^{-t} dt.$$

The recursion relation for the gamma function is $\Gamma(x + 1) = x\Gamma(x)$.

chemical evolution in galaxies. The problem is similarly daunting if the absorbers are *not* associated with galaxies. In either case, we must account for heavy elements in the apparent absence of luminous material (stars).

There is at least one more direct way of measuring the sizes of the absorbing clouds, by examining QSOs that appear close together on the sky (i.e., have similar sight lines) for common absorption systems. This works especially well with the separate images of gravitationally lensed QSOs. Recent determinations of sizes of Lyα clouds on the basis of common absorption systems yield sizes in the range $40h_0^{-1}$ kpc to $\sim 300h_0^{-1}$ kpc at $z \approx 1.8$ (Bechtold *et al.* 1994, Dinshaw *et al.* 1994) and $160h_0^{-1}$ kpc to $\sim 860h_0^{-1}$ kpc at $z \approx 0.7$ (Dinshaw *et al.* 1995), where in each case the lower limit is quite firm. Dinshaw *et al.* (1995) conclude that the characteristic size of the Lyα clouds is $\sim 350h_0^{-1}$ kpc. This may, however, represent more of a measure of a 'correlation length' between Lyα-forest clouds rather than the true sizes of coherent clouds.

12.6   The Intergalactic Medium

12.6.1   The Gunn–Peterson Test

The sharp absorption lines discussed in the previous sections tell us about gas in more or less clumpy forms in the intergalactic medium, but how can we test for the presence of a uniform distribution of neutral hydrogen? The most sensitive test for intergalactic neutral hydrogen is to search for Lyα absorption at $z < z_{em}$. For a smooth distribution, one should look for a depression of the QSO continuum at wavelengths shortward of the Lyα emission line. This first became possible from the ground with the detection of QSOs with $z \gtrsim 2$ in the mid-1960s, and was first carried out by Gunn and Peterson (1965), although it was independently suggested by Scheuer (1965) and in a slightly different form based on Mg II $\lambda2798, 2802$ by Shklovsky (1964).

From the very fact that we can detect continuum shortward of the Lyα emission line, we know that the opacity of neutral hydrogen at large redshift is sufficiently low that we do not need to consider the effects of the damping wings. The broadening of the absorption line due to the cosmological redshift is much larger than that due to thermal motions, so in this context, the width of the Lyα line can be considered to have a profile that can be described by a delta function, i.e.,

$$\sigma(\lambda)\,d\lambda = \frac{\pi e^2}{m_e c}\frac{f\lambda_0^2}{c}\,\delta(\lambda - \lambda_0)\,d\lambda, \tag{12.36}$$

where $\lambda_0 = 1216$ Å and $f = 0.416$ are the wavelength and oscillator strength, respectively, of the Lyα transition. At wavelengths shortward of the redshifted Lyα emission line in some arbitrarily distant quasar at redshift z_{em}, the optical depth at wavelength λ is given by the density of particles and their absorption cross-section at a redshift defined by $(1 + z) = \lambda/\lambda_0$, i.e.,

$$d\tau(\lambda) = n(z)\,\sigma\left(\frac{\lambda}{1+z}\right) c\,dt \tag{12.37}$$

The total optical depth at λ is given by integrating this equation over all redshifts at which neutral hydrogen atoms contribute opacity at λ. With our assumption of a delta-function line profile, it is immediately apparent that along our sight line to a QSO at redshift z_{em}, the optical depth of the intergalactic medium at observed wavelength $\lambda < (1 + z_{em})\lambda_0$ will be due only to those hydrogen atoms at $z = (\lambda/\lambda_0) - 1$, i.e.,

$$
\begin{aligned}
\tau(\lambda) &= \int_0^{z_{em}} n(z)\,\sigma\left(\frac{\lambda}{1+z}\right) c\,\frac{dt}{dz}\,dz \\
&= \frac{\pi e^2}{m_e c}\,\frac{f\lambda_0^2}{c} \times \\
&\quad \int_0^{z_{em}} \frac{c\,n(z)}{H_0(1+z)^2(1+2q_0z)^{1/2}}\,\delta\left(\frac{\lambda}{1+z} - \lambda_0\right)\,dz,
\end{aligned}
\tag{12.38}
$$

where we have used eq. (11.44). The effect of the delta function is to isolate the redshift z such that the condition $(1 + z) = \lambda/\lambda_0$ is met. The integral becomes more transparent by transforming to a variable $x = \lambda/(1 + z)$ (so $dx = -\lambda\,dz/[1 + z]^2$), and the integral over the delta function thus yields a factor $(1 + z)/\lambda_0$. Upon carrying through the integral, the optical depth at $\lambda = (1 + z)\lambda_0$ is

$$
\begin{aligned}
\tau(\lambda) &= \frac{\pi e^2 f\lambda_0}{m_e c}\,\frac{n(z)}{H_0(1+z)(1+2q_0z)^{1/2}} \\
&= \frac{4.14 \times 10^{10}\,n(z)}{h_0(1+z)(1+2q_0z)^{1/2}}.
\end{aligned}
\tag{12.39}
$$

The current observational limit (Steidel and Sargent 1987) is that $\tau < 0.11$ at the 99% confidence level at a mean redshift $\langle z_{abs}\rangle = 2.64$; similar results are found for even higher redshifts (e.g., Giallongo *et al.* 1994). For an Einstein–de Sitter Universe, this yields a limit

$$n(\langle z_{abs}\rangle) < 1.8 \times 10^{-11}\,h_0^{-1}\,\mathrm{cm}^{-3}. \tag{12.40}$$

The present-day density is lower by a factor of $(1 + \langle z_{abs}\rangle)^3 \approx 48$, which gives

$$
\begin{aligned}
\Omega_{H^0} &= \frac{\rho_{H^0}}{\rho_{crit}} = \frac{n(\langle z_{abs}\rangle)\,m_p\,(1+\langle z_{abs}\rangle)^{-3}}{1.9 \times 10^{-29}\,h_0^2} \\
&\approx 3.4 \times 10^{-8}\,h_0^{-3},
\end{aligned}
\tag{12.41}
$$

which is cosmologically insignificant and almost absurdly low. This refers, however, to the density of *neutral* intergalactic hydrogen, and a more sensible conclusion is that the intergalactic medium is ionized. In the next section, we consider whether in fact it is AGNs that ionize the intergalactic medium.

Finally, we mention in passing that in principle the Gunn–Peterson test can also be carried out to test for a highly ionized intergalactic medium by using the He II Lyα $\lambda 304$ line, and at the time of writing a detection of a He II absorption cut-off has been claimed in one object (Jakobsen *et al.* 1994).

12.6.2 Ionization of the Intergalactic Medium

The cosmic microwave background is the detectable remnant of the epoch when the Universe was predominantly ionized and opaque to radiation on account of Thomson scattering by free electrons. As the Universe expanded, the radiation temperature decreased as $T \propto a^{-4}$ (eq. 9.58). At an epoch corresponding to $z \approx 1000$, the radiation temperature had dropped to $\sim 3000\,\text{K}$, at which point the hydrogen recombination rate exceeded the photoionization rate and the Universe became predominantly neutral and transparent. At some later epoch, luminous objects such as AGNs formed and began to emit high-energy radiation that effectively *re-ionized* the intergalactic medium. The question that we will outline briefly is whether the 'metagalactic' flux produced by all AGNs is sufficient to ionize the intergalactic medium to an extent that is sufficient to account for the Gunn–Peterson limits, or whether additional sources of ionization or heating are required. Another way to look at this is to postulate that the integrated high-energy flux from AGNs represents the bulk of the metagalactic ionizing radiation and then calculate an upper limit to the density of the intergalactic medium based on the Gunn–Peterson limit. This turns out to be a complex calculation that depends on many poorly known parameters, and the problem has been treated in the literature by many investigators (see in particular Bechtold *et al.* 1987 and Meiksin and Madau 1993), with no clear consensus emerging.

The specific luminosity of an AGN at some frequency v is taken to be $L_v(v)$ (ergs $\text{s}^{-1}\,\text{Hz}^{-1}$). The total energy radiated at v per unit comoving volume by all AGNs is thus $L_v(v)\,\phi(L,z)$ (ergs $\text{s}^{-1}\,\text{cm}^{-3}\,\text{Hz}^{-1}$), where $\phi(L,z)$ is the QSO luminosity function, and we have assumed for simplicity that L_v is not dependent on z. The quantity we wish to compute is the *proper volume* emissivity of all QSOs as a function of v and z, which can be written

$$\epsilon(v,z) = \int_0^\infty (1+z)^3 L_v(v)\,\phi(L,z)\,dL, \tag{12.42}$$

where the $(1+z)^3$ factor adjusts from comoving to proper density.

We now consider the mean specific intensity of the AGN radiation field $J_v(v_0, z_0)$ (ergs $\text{s}^{-1}\,\text{cm}^{-2}\,\text{Hz}^{-1}\,\text{ster}^{-1}$)†, at some particular frequency v_0 and redshift z_0, which is due to the integrated light from all AGNs at $z > z_0$; we recall that the radiation received at frequency v_0 at the epoch corresponding to redshift z_0 was emitted at a frequency

$$v(z) = \left(\frac{1+z}{1+z_0}\right) v_0 \tag{12.43}$$

at redshift z. Of course, the light arriving from some AGN at a higher redshift z will be

---

† The mean specific intensity is the specific intensity averaged over all directions, i.e.,

$$J_v = \frac{1}{4\pi} \int\int I_v(\theta, \phi)\, \sin\theta\, d\theta\, d\phi.$$

attenuated by some amount $\tau(v_0, z_0, z)$, whose form will depend on the specific model of the intergalactic medium. For example, if the principal source of intergalactic opacity for H-ionizing photons is Lyα clouds, the mean opacity, averaged over all directions, will be

$$\tau_{\text{eff}} = \int_{z_0}^{z} \int_{0}^{\infty} \frac{\partial^2 N}{\partial N(\text{H\,I})\, \partial z'} \times$$
$$\{1 - \exp\left[-\tau_{\text{cloud}}(N(\text{H\,I}), v_0, z')\right]\}\, dN(\text{H\,I})\, dz'. \qquad (12.44)$$

This formula applies in the case of a clumpy medium which contains clouds of individual optical depth τ_{cloud} (Paresce, McKee, and Bowyer 1980), where in this case

$$\tau_{\text{cloud}}(N(\text{H\,I}), v_0, z) = N(\text{H\,I})\, \sigma \left(\frac{v_0(1+z)}{1+z_0} \right). \qquad (12.45)$$

The cross-section σ is the cross-section for photoionization of hydrogen evaluated at a frequency given by eq. (12.43). Additional sources of opacity, such as a smoothly distributed component of the intergalactic medium or dust (Heisler and Ostriker 1988) might also cause attenuation of the ionizing flux from distant AGNs.

With these various functions in hand, we can compute the mean specific intensity of AGN light at v_0 and z_0, noting again that the proper-length interval $d\ell/dz = c\, dt/dz$ is given by eq. (11.44),

$$J_v(v_0, z_0) = \int_{z_0}^{\infty} \left(\frac{1+z_0}{1+z} \right)^3 \epsilon(v(z), z) \times$$
$$\exp\left[-\tau_{\text{eff}}(v(z), z_0, z)\right] \frac{d\ell}{dz} dz$$
$$= \frac{c}{4\pi H_0} \int_{z_0}^{\infty} \left(\frac{1+z_0}{1+z} \right)^3 \epsilon(v(z), z) \times$$
$$\frac{\exp\left[-\tau_{\text{eff}}\right] dz}{(1+z)^2(1+2q_0 z)^{1/2}}. \qquad (12.46)$$

It should be noted that $\epsilon \propto L\, \phi(L, z)$ and since $L \propto h_0^{-2}$ and $\phi \propto h_0^3$, J_v is *independent* of H_0, though it does depend on q_0.

As mentioned earlier, evaluation of J_v is difficult and subject to uncertainties in the QSO luminosity function, the intrinsic shape of AGN spectra at H-ionizing frequencies, and the opacity of the Universe. Despite these uncertainties, the results obtained by various investigators are surprisingly consistent, with the mean intensity at the Lyman edge v_1 typically having a value

$$J(v_1, z) \approx 3 \times 10^{-22}\ \text{ergs s}^{-1}\,\text{cm}^{-2}\,\text{Hz}^{-1}\,\text{ster}^{-1} \qquad (12.47)$$

at $z \approx 3$.

There is an independent way that J_v can be measured directly, namely through the proximity effect in the Lyα forest (Bajtlik, Duncan, and Ostriker 1988). The basic idea is to determine the limiting redshift $z_L < z_{\text{em}}$ where the proximity effect vanishes in a particular QSO spectrum; through the Hubble law, this can be translated into a distance r_L where the condition

$$\frac{L_v}{4\pi r_L^2} \approx 4\pi J_v \qquad (12.48)$$

is met. In other words, the proximity effect vanishes at a distance r_L from the QSO where the ionizing flux from the nearby QSO is equal to the integrated background flux from all QSOs. This simple treatment is valid as long as the spectral shapes of the QSO and the background radiation (which is also due to QSOs) are identical. Again, the value of J_v so derived is independent of H_0, and is consistent with the values found by solution of eq. (12.46), i.e.,

$$J(v_1, z) = 10^{-21.0 \pm 0.5} \text{ ergs s}^{-1} \text{ cm}^{-2} \text{ Hz}^{-1} \text{ ster}^{-1} \qquad (12.49)$$

for $1.7 < z < 3.8$ (Bajtlik, Duncan, and Ostriker 1988). The implicit uncertainties in this method are also numerous: the assumption of ionization equilibrium may not be valid if the QSO is variable and the recombination time (eq. 5.33) for the absorbing clouds is long. Also, the assumption that the space density of the gas clouds that produce the Lyα forest is the same in the immediate environment of QSOs as it is everywhere else is likely to be incorrect. Nevertheless the order-of-magnitude consistency of the two independent methods is reassuring.

Finally, we also mention that J_v at $z \approx 0$ can be inferred indirectly from Hα-emission measurements of optically thick local intergalactic clouds. Vogel *et al.* (1995) deduce an upper limit $J_v(v_1, z = 0) < 8 \times 10^{-23} \text{ ergs s}^{-1} \text{ cm}^{-2} \text{ Hz}^{-1} \text{ ster}^{-1}$; the value expected from application of eq. (12.46) at $z = 0$ is only about an order of magnitude below this limit.

Once the background ionizing flux has been established, the Gunn–Peterson limit can be used to place an upper limit on the density of a smoothly distributed extragalactic medium. If the density is too high, the measured flux will be unable to keep the medium ionized and a Gunn–Peterson absorption trough should be observable. Current estimates of J_v yield an upper limit

$$\Omega_{\text{IGM}} \lesssim 0.08 h_0^{-2}, \qquad (12.50)$$

(Meiksin and Madau 1993). If independent evidence for a higher value of Ω_{IGM} were found, we would be forced to conclude that the ionizing background at higher z must be larger than we can account for with AGNs alone. Another source of ionizing radiation, perhaps massive starbursts at high redshift, would have to be invoked to explain the high level of ionization of the intergalactic medium.

References

Abell, G.O. 1962, in *Problems of Extra-Galactic Research*, ed. G.C. McVittie (Macmillan: New York), p. 213

Adams, T.F. 1977, ApJS, 33, 19

Aldcroft, T.L., Bechtold, J., and Elvis, M. 1994, ApJS, 93, 1

Alloin, D., *et al.* 1995, A&A, 293, 293

Antonucci, R. 1993, ARAA, 31, 473

Antonucci, R., Hurt, T., and Miller, J. 1994, ApJ, 430, 210

Antonucci, R.R.J., Kinney, A.L., and Ford, H.C. 1989, ApJ, 342, 64

Antonucci, R.R.J., and Miller, J.S. 1985, ApJ, 297, 621

Arnaud, K.A., Branduardi-Raymont, G., Culhane, J.L., Fabian, A.C., Hazard, C., McGlynn, T.A., Shafer, R.A., Tennant, A.F., and Ward, M.J. 1985, MNRAS, 217, 105

Atwood, B., Baldwin, J.A., and Carswell, R.F. 1982, ApJ, 257, 559

Avni, Y., and Bahcall, J.N. 1980, ApJ, 235, 694

Baade, W., and Minkowski, R. 1954, ApJ, 119, 215

Bahcall, J.N., Kirhakos, S., and Schneider, D.P. 1995, ApJ, 450, 486

Bahcall, J.N., Jannuzi, B.T., Schneider, D.P., Hartig, G.F., Bohlin, R., Junkkarinen, V. 1991, ApJ, 377, L5

Bahcall, J.N. *et al.* 1993, ApJS, 87, 1

Bajtlik, S., Duncan, R.C., and Ostriker, J.P. 1988, ApJ, 327, 570

Baker, J.G., and Menzel, D.H. 1938, ApJ, 88, 52

Baldwin, J.A. 1977, ApJ, 214, 679

Baldwin, J., Ferland, G., Korista, K., and Verner, D. 1995, ApJ, 455, L119

Baldwin, J.A., Phillips, M.M., and Terlevich, R. 1981, PASP, 93, 5

Barlow, T.A., Junkkarinen, V.T., Burbidge, E.M., Weymann, R.J., Morris, S.L., and Korista, K.T. 1992, ApJ, 397, 81

Barnes, J.E., and Hernquist, L.E. 1991, ApJ, 370, L65

Barthel, P.D. 1989, ApJ, 336, 606

Barvainis, R. 1987, ApJ, 320, 537

Barvainis, R. 1993, ApJ, 412, 513

Baum, S.A., and Heckman, T. 1989, ApJ, 336, 702

Bechtold, J. 1994, ApJS, 91, 1

Bechtold, J., Crotts, A.P.S., Duncan, R.C., and Fang, Y. 1994, ApJ, 437, L83

Bechtold, J., Weymann, R.J., Lin, Z., and Malkan, M.A. 1987, ApJ, 315, 180

Begelman, M.C. 1985, in *Astrophysics of Active Galaxies and Quasi-Stellar Objects*, ed. J.S. Miller (University Science Books: Mill Valley), p. 411

Bennett, A.S. 1961, MemRAS, 68, 163

Bergeron, J., *et al.* 1994, ApJ, 436, 33

Blaauw, A., and Elvius, T. 1965, in *Stars and Stellar Systems: Galactic Structure*, ed. A. Blaauw and M. Schmidt (University of Chicago Press: Chicago), p. 589

Black, J.H., Chaffee, F.H., Jr., and Foltz, C.B. 1987, ApJ, 317, 442

Blandford, R.D. 1985, in *Active Galactic Nuclei*, ed. J.E. Dyson, (Manchester Univ. Press: Manchester), p. 281

Blandford, R.D. 1990, in *Active Galactic Nuclei*, ed. T.J.-L. Courvoisier and M. Mayor, (Springer-Verlag: Berlin), p. 161

Blandford, R.D., and McKee, C.F. 1982, ApJ, 255, 419

Blandford, R.D., McKee, C.F., and Rees, M.J. 1977, Nature, 267, 211

Blandford, R.D., and Rees, M.J. 1978, in *Pittsburgh Conference on BL Lac Objects*, ed. A.M. Wolfe (University of Pittsburgh: Pittsburgh), p. 328

Boroson, T.A., and Green, R.F. 1992, ApJS, 80, 109

Boroson, T.A., and Oke, J.B. 1982, Nature, 296, 397

Boroson, T.A., Oke, J.B., and Green, R.F. 1982, ApJ, 263, 32

Boyle, B.J., Jones, L.R., Shanks, T., Marano, B., Zitelli, V., and Zamorani, G. 1991, in *The Space Distribution of Quasars*, ed. D. Crampton (Astronomical Society of the Pacific: San Francisco), p. 191

Bregman, J.N. 1990, A&A Rev., 2, 125

Bregman, J.N., *et al.* 1986, ApJ, 301, 708

Bregman, J.N., *et al.* 1990, ApJ, 352, 574

Bridle, A.H., Hough, D.H., Lonsdale, C.J., Burns, J.O., and Laing, R.A. 1994, AJ, 108, 766

Bridle, A.H., and Perley, R.A. 1984, ARAA, 22, 319

Burbidge, G., and Burbidge, M. 1967, *Quasi-Stellar Objects* (W.H. Freeman: San Francisco)

Burbidge, G.R., Burbidge, E.M., and Sandage, A.R. 1963, Rev. Mod. Phys., 35, 947.

Burbidge, G., and Hewitt, A. 1992, in *Variability of Blazars*, ed. E. Valtaoja and M. Valtonen (Cambridge University Press: Cambridge), p. 4

Burstein, D., and Heiles, C. 1982, AJ, 87, 1165

Burstein, D., and Heiles, C. 1984, ApJS, 54, 33

Cai, W., and Pradhan, A.K. 1993, ApJS, 88, 329

Capetti, A., Macchetto, F., Axon, D.J., Sparks, W.B., and Boksenberg, A. 1995, ApJ, 452, L87

Capriotti, E., Foltz, C., and Byard, P. 1980, ApJ, 241, 903

Capriotti, E., Foltz, C., and Byard, P. 1981, ApJ, 245, 396

Cardelli, J.A., Clayton, G.C., and Mathis, J.S. 1989, ApJ, 345, 245

Carswell, R.F. 1989, in *The Epoch of Galaxy Formation*, ed. C.S. Frenk, R.S. Ellis, T. Shanks, A.F. Heavens, and J.A. Peacock (Kluwer Academic Publishers: Dordrecht), p. 89

Chaffee, F.H., Jr., Black, J.H., and Foltz, C.B. 1988, ApJ, 335, 584

Cheng, F.-Z., Danese, L., De Zotti, G., and Franceschini, A. 1985, MNRAS, 212, 857

Chini, R., Kreysa, E., and Biermann, P.L. 1989, A&A, 219, 87

Chiu, H.-Y. 1964, Phys. Today, 17, No. 5, 21

Clavel, J., Wamsteker, W., and Glass, I.S. 1989, ApJ, 337, 236

Clavel, J., *et al.* 1991, ApJ, 366, 64

Code, A.D., *et al.* 1993, ApJ, 403, L63

Cohen, M.H. 1989, in *BL Lac Objects*, ed. L. Maraschi, T. Maccacaro, and M.-H. Ulrich, (Springer-Verlag: Berlin), p. 13

Cohen, R.D. 1983, ApJ, 273, 489

Condon, J.J. 1984, ApJ, 287, 461

Corbin, M.R. 1990, ApJ, 357, 346

Crenshaw, D.M. 1986, ApJS, 62, 821

Curtis, H.D. 1913, Lick Obs. Publ., 13, 11

Czerny, B., and Elvis, M. 1987, ApJ, 321, 305

Dahari, O. 1984, AJ, 89, 966

Daly, R.A. 1990, ApJ, 355, 416

De Robertis, M.M., and Osterbrock, D.E. 1984, ApJ, 286, 171

De Robertis, M.M., and Osterbrock, D.E. 1986, ApJ, 301, 727

de Vaucouleurs, G. 1953, MNRAS, 113, 134

De Zotti, G., and Gaskell, C.M. 1985, A&A, 147, 1

Dinshaw, N., Foltz, C.B., Impey, C.D., Weymann, R.J., and Morris, S.L. 1995, Nature, 373, 223

Dinshaw, N., Impey, C.D., Foltz, C.B., Weymann, R.J., and Chaffee, F.H. 1994, ApJ, 437, L87

Done, C., and Fabian, A.C. 1989, MNRAS, 240, 81

Draine, B.T., and Lee, H.M. 1984, ApJ, 285, 89

Dressler, A., Thompson, I.B., and Shectman, S.A. 1985, ApJ, 288, 481

Eddington, A.S. 1913, MNRAS, 73, 359

Edelson, R. 1992, ApJ, 401, 516

Edelson, R.A., Gear, W.K.P., Malkan, M.A., and Robson, E.I. 1988, Nature, 336, 749

Edelson, R.A., and Malkan, M.A. 1987, ApJ, 323, 516

Edelson, R., et al. 1995, ApJ, 438, 120

Edge, D.O., Shakeshaft, J.R., McAdam, W.B., Baldwin, J.E., and Archer, S. 1959, MemRAS, 68, 37

Ehman, J.R., Dixon, R.S., and Kraus, J.D. 1970, AJ, 75, 351

Ekers, J.A. 1969, Austral. J. Phys. Astrophys. Suppl. No. 7, 1

Ellingson, E., Yee, H.K.C., and Green, R.F. 1991, ApJ, 371, 49

Elvis, M., Fassnacht, C., Wilson, A.S., and Briel, U. 1990, ApJ, 361, 459

Elvis, M., Lockman, F.J., Wilkes, B.J. 1989, AJ, 97, 777

Elvis, M., Maccacaro, T., Wilson, A.S., Ward, M.J., Penston, M.V., Fosbury, R.A.E., Perola, G.C. 1978, MNRAS, 183, 129

Elvis, M., Wilkes, B.J., McDowell, J.C., Green, R.F., Bechtold, J., Willner, S.P., Oey, M.S., Polomski, E., and Cutri, R. 1994, ApJS, 95, 1

Espey, B.R., et al. 1994, ApJ, 434, 484

Faber, S.M., and Jackson, R.E. 1976, ApJ, 204, 668

Fall, S.M., and Pei, Y.C. 1993, ApJ, 402, 479

Fanaroff, B.L., and Riley, J.M. 1974, MNRAS, 167, 31P

Felten, J.E. 1977, AJ, 82, 861

Ferland, G.J., Korista, K.T., and Peterson, B.M. 1990, ApJ, 363, L21

Ferland, G.J., and Osterbrock, D.E. 1986, ApJ, 300, 658

Ferland, G.J., Peterson, B.M., Horne, K., Welsh, W.F., and Nahar, S.N. 1992, ApJ, 387, 95

Filippenko, A.V. 1992, in *Relationships Between Active Galactic Nuclei and Starburst Galaxies*, ed. A.V. Filippenko (Astronomical Society of the Pacific: San Francisco), p. 253

Filippenko, A.V., and Halpern, J.P. 1984, ApJ, 285, 458

Filippenko, A.V., Ho, L.C., and Sargent, W.L.W. 1993, ApJ, 410, L75

Fitch, W.S., Pacholczyk, A.G., and Weymann, R.J. 1967, ApJ, 150, L67

Foltz, C.B., Chaffee, F.H., Jr., and Black, J.H. 1988, ApJ, 324, 267

Foltz, C.B., Chaffee, F.H., Jr., Hewett, P.C., MacAlpine, G.M., Turnshek, D.A., Weymann, R.J., and Anderson, S.F. 1987, AJ, 94, 1423

Foltz, C.B., Chaffee, F.H., Hewett, P.C., Weymann, R.J., and Morris, S.L. 1990, BAAS, 22, 806

Foltz, C.B., Weymann, R.J., Peterson, B.M., Sun, L., Malkan, M.A., and Chaffee, F.H., Jr. 1986, ApJ, 307, 504

Ford, H.C., Harms, R.J., Tsvetanov, Z.I., Hartig, G.F., Dressel, L.L., Kriss, G.A., Bohlin, R.C., Davidsen, A.F., Margon, B., and Kochhar, A.K. 1994, ApJ, 435, L27

Francis, P.J., Hewett, P.C., Foltz, C.B., Chaffee, F.H., Weymann, R.J., and Morris, S.L. 1991, ApJ, 373, 465

Frank, J., King, A.R., and Raine, D.J. 1992, *Accretion Power in Astrophysics*, 2nd Edition (Cambridge University Press: Cambridge)

Freeman, K.C. 1970, ApJ, 160, 811

Garrington, S.T., Leahy, J.P., Conway, R.G., and Laing, R.A. 1988, Nature, 331, 147

Gaskell, C.M. 1982, ApJ, 263, 79

Gaskell, C.M., and Ferland, G.J. 1984, PASP, 96, 393

Gehren, T., Fried, J., Wehinger, P.A., and Wyckoff, S. 1984, ApJ, 278, 11

George, I.M., and Fabian, A.C. 1991, MNRAS, 249, 352

Giallongo, E., D'Odorico, S., Fontana, A., McMahon, R.G., Savaglio, S. Cristiani, S., Molaro, P., Trevese, D. 1994, ApJ, 425, L1

Gioia, I.M., Maccacaro, T., Schild, R.E., Wolter, A., Stocke, J.T., Morris, S.L., and Henry, J.P. 1990, ApJS, 72, 567

Goodman, J. 1995, Phys. Rev. D, 52, 1821

Gondhalekar, P.M., Kellet, B.J., Pounds, K.A., Matthews, L., and Quenby, J.J. 1994, MNRAS, 268, 973

Gower, J.F.R., Scott, P.F., and Wills, D. 1967, MemRAS, 71, 49

Greenstein, J.L., and Matthews, T.A. 1963, Nature, 197, 1041

Guilbert, P.W., Fabian, A.C., and Rees, M.J. 1983, MNRAS, 205, 593

Guilbert, P.W., and Rees, M.J. 1988, MNRAS, 233, 475

Gunn, J.E. 1979, in *Active Galactic Nuclei* (Cambridge University Press: Cambridge), p. 213

Gunn, J.E., and Peterson, B.A. 1965, ApJ, 142, 1633

Haardt, F., and Maraschi, L. 1993, ApJ, 413, 507

Halpern, J.P. 1984, ApJ, 281, 90

Hamann, F., Korista, K.T., and Morris, S.L. 1993, ApJ, 415, 541

Haro, G., and Luyten, W.J. 1962, Bol. Obs. Tonantzintla y Tacubaya, 2, No. 16, 3

Hartwick, F.D.A., and Schade, D. 1990, ARAA, 28, 437

Hawkins, M.R.S. 1986, MNRAS, 219, 417

Hazard, C., Gulkis, S., and Bray, A.D. 1967, ApJ, 148, 669

Hazard, C., Mackey, M.B., and Shimmins, A.J. 1963, Nature, 197, 1037

Heckman, T.M. 1978, PASP, 90, 241

Heckman, T.M. 1980, A&A, 87, 152

Heckman, T.M. 1987, in *Observational Evidence of Activity in Galaxies*, ed. E.Ye. Khachikian, K.J. Fricke, and J. Melnick (Reidel: Dordrecht), p. 421

Heckman, T.M., Bothun, G.D., Balick, B., and Smith, E.P. 1984, AJ, 89, 958

Heckman, T.M., Balick, B., and Sullivan, W.T., III 1978, ApJ, 224, 745

Heckman, T.M., Chambers, K.C., and Postman, M. 1992, ApJ, 391, 39

Heisler, J., and Ostriker, J.P. 1988, ApJ, 332, 543

Hewett, P. 1992, in *Physics of Active Galactic Nuclei*, ed. W.J. Duschl and S.J. Wagner (Springer-Verlag: Berlin), p. 649

Hewett, P.C., and Foltz, C.B. 1994, PASP, 106, 113

Hewett, P.C., Foltz, C.B., and Chaffee, F.H. 1995, AJ, 109, 1498

Hewett, P.C., Irwin, M.J., Bunclark, P., Bridgeland, M.T., Kibblewhite, E.J., He, X.-T., and Smith, M.G. 1985, MNRAS, 213, 971

Hewitt, A., and Burbidge, G. 1993, ApJS, 87, 451

Ho, L., Filippenko, A.V., and Sargent, W.L.W. 1993, ApJ, 417, 63

Ho, L., Filippenko, A.V., and Sargent, W.L.W. 1994, in *Multi-Wavelength Continuum Emission of AGN*, ed. T.J.-L. Courvoisier and A. Blecha (Kluwer Academic Publishers: Dordrecht), p. 275

Ho, L., Filippenko, A.V., and Sargent, W.L.W. 1995, ApJS, 98, 477

Holmberg, E. 1975, in *Galaxies and the Universe*, ed. A. Sandage, M. Sandage, and J. Kristian (University of Chicago Press: Chicago), p. 123

Hook, I.M., McMahon, R.G., Boyle, B.J., and Irwin, M.J. 1994, MNRAS, 268, 305

Huchra, J.P. 1977, ApJS, 35, 171

Hughes, D.H., Robson, E.I., Dunlop, J.S., and Gear, W.K. 1993, MNRAS, 263, 607

Hummer, D.G. 1970, MemRAS, 70, 1

Hutchings, J.B. 1987, ApJ, 320, 122

Hutchings, J.B., and Crampton, D. 1990, AJ, 99, 37

Hutchings, J.B., Crampton, D., and Campbell, B. 1984, ApJ, 280, 41

Hutchings, J.B., Janson, T., and Neff, S.G. 1989, ApJ, 342, 660

Jackson, N. and Browne, I.W.A. 1990, Nature, 343, 43

Jackson, N. and Browne, I.W.A. 1991, MNRAS, 250, 422

Jakobsen, P., Boksenberg, A., Deharveng, J.M., Greenfield, P., Jedrzejewski, R., and Paresce, F. 1994, Nature, 370, 35

Jannuzi, B.T., Elston, R., Schmidt, G.D., Smith, P.S., and Stockman, H.S. 1994, ApJ, 429, L49

Johnson, W.N., *et al.* 1994, in *The Second Compton Symposium* (American Institute of Physics: New York), p. 515

Jones, T.W., O'Dell, S.L., and Stein, W.A. 1974, ApJ, 188, 353

Kazanas, D. 1989, ApJ, 347, 74

Kellermann, K.I., and Pauliny-Toth, I.I.K. 1969, ApJ, 155, L71

Kellermann, K.I., Sramek, R., Schmidt, M., Shaffer, D.B., and Green, R. 1989, AJ, 98, 1195

Kennefick, J.D., Djorgovski, S.G., and de Carvalho, R.R. 1995, AJ, 110, 2553

Kerrick, A.D., *et al.* 1995, ApJ, 449, L99

Khachikian, E.Ye., and Weedman, D.W. 1974, ApJ, 192, 581

Kinney, A.L., Antonucci, R.R.J., Ward, M.J., Wilson, A.S., and Whittle, M. 1991, ApJ, 377, 100

Kinney, A.L., Rivolo, A.R., and Koratkar, A.P. 1990, ApJ, 357, 338

Kjærgaard, P. 1978, Physica Scripta, 17, 347

Kollatschny, W., and Fricke, K.J. 1989, A&A, 219, 34

Kollatschny, W., Netzer, H., and Fricke, K.J. 1986, A&A, 163, 31

Koo, D.C., Kron, R.G., and Cudworth, K.M. 1986, PASP, 98, 285

Koratkar, A.P., Kinney, A.L., Bohlin, R.C. 1992, ApJ, 400, 435

Koratkar, A., Antonucci, R.R.J., Goodrich, R.W., Bushouse, J., and Kinney, A.L. 1995, ApJ, 450, 501

Korista, K.T., Voit, G.M., Morris, S.L., and Weymann, R.J. 1993, ApJS, 88, 357

Korista, K.T., *et al.* 1995, ApJS, 97, 285

Kormendy, J., and Richstone, D. 1995, ARAA, 33, 581

Koski, A.T. 1978, ApJ, 223, 56

Kotilainen, J.K., and Ward, M.J. 1994, MNRAS, 266, 953

Kraus, J.D. 1966, *Radio Astronomy* (McGraw-Hill: New York), p. 439

Kristian, J. 1973, ApJ, 179, L61

Krolik, J.H., and Begelman, M.C. 1988, ApJ, 329, 702

Krolik, J.H., and Lepp, S. 1989, ApJ, 347, 179

Krolik, J.H., Madau, P., and Życki, P.T. 1994, ApJ, 420, L57

Krolik, J.H., McKee, C.F., and Tarter, C.B. 1981, ApJ, 249, 422

Kron, R.G., and Chiu, L.-T.G. 1981, PASP, 93, 397

Kwan, J., and Krolik, J.H. 1981, ApJ, 250, 478

Laing, R.A. 1988, Nature, 331, 149

Laing, R.A., and Bridle, A.H. 1987, MNRAS, 228, 557

Lanzetta, K.M., Wolfe, A.M., and Turnshek, D.A. 1995, ApJ, 440, 435

Laor, A., and Draine, B.T. 1993, ApJ, 402, 441

Laor, A., and Netzer, H. 1989, MNRAS, 238, 897

Laor, A., Netzer, H., and Piran, T. 1990, MNRAS, 242, 560

Lawrence, A. 1991, MNRAS, 252, 586

Lawrence, A., and Elvis, M. 1982, ApJ, 256, 410

Lebofsky, M.J., and Rieke, G.H. 1980, Nature, 284, 410

Lemaître, G. 1931, MNRAS, 91, 483

Levshakov, S.A., and Varshalovich, D.A. 1985, MNRAS, 212, 517

Lightman, A.P., and White, T.R. 1988, ApJ, 335, 57

Lipovetsky, V.A., Markarian, B.E., and Stepanian, J.A. 1987, in *Observational Evidence of Activity in Galaxies*, ed. E. Ye. Khachikian, K.J. Fricke, and J. Melnick (Dordrecht: Reidel), p. 17

Lockman, F.J., and Savage, B.D. 1995, ApJS, 97, 1

Longair, M.S. 1978, in *Observational Cosmology*, ed. A. Maeder, L. Martinet, and G. Tammann (Geneva Observatory: Geneva), p. 127

Loveday, J., Peterson, B.A., Efstathiou, G., Maddox, S.J. 1992, ApJ, 390, 338

Low, F.J., Huchra, J.P., Kleinmann, S.G., and Cutri, R.M. 1988, ApJ, 327, L41

Lynds, C.R. 1967, ApJ, 147, 396

Lynds, R. 1971, ApJ, 164, L73

MacAlpine, G.M. 1985, in *Astrophysics of Active Galaxies and Quasi-Stellar Objects*, ed. J.S. Miller (University Science Books: Mill Valley), p. 259

MacKenty, J.W. 1990, ApJS, 72, 231

Madejski, G.M., Done, C., Turner, T.J., Mushotzky, R.F., Serlemitsos, P., Fiore, F., Sikora, M., and Begelman, M.C. 1993, Nature, 365, 626

Maisack, M., Johnson, W.N., Kinzer, R.L., Strickman, M.S., Kurfess, J.D., Cameron, R.A., Jung, G.V., Grabelsky, D.A., Purcell, W.R., and Ulmer, M.P. 1993, ApJ, 407, L61

Maisack, M., *et al.* 1995, A&A, 298, 400

Makino, F., *et al.* 1987, ApJ, 313, 662

Malkan, M.A. 1984, ApJ, 287, 555

Malkan, M.A. 1989, in *Theory of Accretion Disks*, ed. F. Meyer, W.J. Duschl, J. Frank, and E. Meyer-Hofmeister (Kluwer Academic Publishers: Dordrecht), p. 19

Malkan, M.A., Margon, B., and Chanan, G.A. 1984, ApJ, 280, 66

Malkan, M.A., and Sargent, W.L.W. 1982, ApJ, 254, 22

Marshall, H.L. 1987, AJ, 94, 628

Marzke, R.O., Huchra, J.P., and Geller, M.J. 1994, ApJ, 428, 43

Mathews, W.G., and Capriotti, E.R. 1985, in *Astrophysics of Active Galaxies and Quasi-Stellar Objects*, ed. J.S. Miller (University Science Books: Mill Valley), p. 185

Mathews, W.G., and Ferland, G.J. 1987, ApJ, 323, 456

Mathur, S., Elvis, M., and Wilkes, B. 1995, ApJ, 452, 230

Matthews, T.A., and Sandage, A.R. 1963, ApJ, 138, 30

McCarthy, P.J., Dickinson, M., Filippenko, A.V., Spinrad, H., and van Breugel, W.J.M. 1988, ApJ, 328, L29

McHardy, I. 1988, Mem. Soc. Astron. Ital., 59, 239

McHardy, I. 1990, in *Proceedings of the 23rd ESLAB Symposium*, ed. J. Hunt and B. Battrick (ESA: Paris), p. 1111

McLeod, K.K., and Rieke, G.H. 1995a, ApJ, 441, 96

McLeod, K.K., and Rieke, G.H. 1995b, ApJ, 454, L77

Meiksin, A., and Madau, P. 1993, ApJ, 412, 34

Mihalas, D., and Binney, J. 1981, *Galactic Astronomy: Structure and Kinematics* (W.H. Freeman: San Francisco)

Miley, G.K., and Miller, J.S. 1979, ApJ, 228, L55

Miller, H.R., Carini, M.T., and Goodrich, B.D. 1989, Nature, 337, 627

Miller, J.S. 1989, in *BL Lac Objects*, ed. L. Maraschi, T. Maccacaro, and M.-H. Ulrich (Springer-Verlag: Berlin), p. 395

Miller, J.S., and Goodrich, R.W. 1990, ApJ, 355, 456

Miller, J.S., and Goodrich, R.W., and Mathews, W.G. 1991, ApJ, 378, 47

Minkowski, R. 1957, in *Radio Astronomy*, ed. H.C. Van de Hulst, (Cambridge University Press: Cambridge), p. 107

Mirabel, I.F., and Wilson, A.S. 1984, ApJ, 277, 92

Mittaz, J.P.D., and Branduardi-Raymont, G. 1989, MNRAS, 238, 1029

Miyoshi, M., Moran, J., Herrnstein, J., Greenhill, L., Nakai, N., Diamond, P., and Inoue, M. 1995, Nature, 373, 127

Morgan, W.W. 1958, PASP, 70, 364

Morris, S.L. and Ward, M.J. 1989, ApJ, 340, 713

Morris, S.L., Weymann, R.J., Savage, B.D., and Gilliland, R.L. 1991, ApJ, 377, L21

Murdoch, H.S., Hunstead, R.W., Pettini, M., and Blades, J.C. 1986, ApJ, 309, 19

Mushotzky, R.F., Done, C., and Pounds, K.A. 1993, ARAA, 31, 717

Mushotzky, R.F., Marshall, F.E., Boldt, E.A., Holt, S.S., and Serlemitsos, P.J. 1980, ApJ, 235, 377

Mushotzky, R.F., *et al.* 1995, MNRAS, 272, L9

Nandra, K., and Pounds, K.A. 1994, MNRAS, 268, 405

Netzer, H., and Laor, A. 1993, ApJ, 404, L51

Norman, C., and Scoville, N. 1988, ApJ, 332, 124

Oke, J.B., and Sandage, A. 1968, ApJ, 154, 21

Orr, M.J.L., and Browne, I.W.A. 1982, MNRAS, 200, 1067

Osmer, P.S. 1981, ApJ, 247, 762

Osmer, P.S. 1982, ApJ, 253, 28

Osterbrock, D.E. 1977, ApJ, 215, 733

Osterbrock, D.E. 1978, Proc. Natl. Acad. Sci. USA, 75, 540

Osterbrock, D.E. 1981, ApJ, 249, 462

Osterbrock, D.E. 1989, *Astrophysics of Gaseous Nebulae and Active Galactic Nuclei* (University Science Books: Mill Valley)

Padovani, P. 1993, MNRAS, 263, 461

Padovani, P., Burg, R., and Edelson, R.A. 1990, ApJ, 353, 438

Padovani, P., and Urry, C.M. 1992, ApJ, 387, 449

Paresce, F., McKee, C.F., and Bowyer, S. 1980, ApJ, 240, 387

Patnaik, A.R., Browne, I.W.A., Walsh, D., Chaffee, F.H., and Foltz, C.B. 1992, MNRAS, 259, 1p

Pearson, T.J., Unwin, S.C., Cohen, M.H., Linfield, R.P., Readhead, A.C.S., Seielstad, G.A., Simon, R.S., and Walker, R.C. 1981, Nature, 290, 365

Pedlar, A., Dyson, J.E., and Unger, S.W. 1985, MNRAS, 214, 463

Peebles, P.J.E. 1971, *Physical Cosmology* (Princeton University Press: Princeton)

Peebles, P.J.E. 1993, *Principles of Physical Cosmology* (Princeton University Press: Princeton)

Pei, Y.C., Fall, S.M., and Bechtold, J. 1991, ApJ, 378, 6

Pelat, D., Alloin, D., and Fosbury, R.A.E. 1981, MNRAS, 195, 787

Penston, M.V. 1988, MNRAS, 233, 601

Penston, M.V., and Cannon, R.D. 1970, R. Obs. Bull., No. 159, p. 84

Penston, M.V., and Pérez, E. 1984, MNRAS, 211, 33p

Peterson, B.A. 1988, in *Proceedings of a Workshop on Optical Surveys for Quasars*, ed. P.S. Osmer, A.C. Porter, R.F. Green, and C.B. Foltz (Astronomical Society of the Pacific: San Francisco), p. 23

Peterson, B.M. 1993, PASP, 105, 247

Peterson, B.M. 1994, in *Reverberation Mapping of the Broad-Line Region in Active Galactic Nuclei*, ed. P.M. Gondhalekar, K. Horne, and B.M. Peterson (Astronomical Society of the Pacific: San Francisco), p. 1

Peterson, B.M., *et al.* 1991, ApJ, 368, 119

Petitjean, P., Webb, J.K., Rauch, M., Carswell, R.F., and Lanzetta, K. 1993, MNRAS, 262, 499

Petrosian, A.R., Saakian, K.A., and Khachikian, E.Ye. 1979, Astrophys., 15, 250

Pettini, M., Smith, L.J., Hunstead, R.W., and King, D.L. 1994, ApJ, 426, 79

Phinney, E.S. 1985, in *Astrophysics of Active Galaxies and Quasi-Stellar Objects*, ed. J.S. Miller (University Science Books: Mill Valley), p. 453

Pier, E.A., and Krolik, J.H. 1992, ApJ, 401, 99

Pilkington, J.D.H., and Scott, P.F. 1965, MemRAS, 69, 183

Pogge, R.W. 1988, ApJ, 328, 519

Pogge, R.W. 1992, in *Astronomical CCD Observing and Reduction Techniques*, ed. S.B. Howell (Astronomical Society of the Pacific: San Francisco), p. 195

Pogge, R.W., and De Robertis, M.M. 1993, ApJ, 404, 563

Pogge, R.W., and Peterson, B.M. 1992, AJ, 103, 1084

Press, W.H., and Rybicki, G.B. 1993, ApJ, 418, 585

Prieto, M.A., and Freudling, W. 1993, ApJ, 418, 668

Pritchet, C.J., and Hartwick, F.D.A. 1987, ApJ, 320, 464

Punch, M., *et al.* 1992, Nature, 358, 477

Quirrenbach, A., *et al.* 1991, ApJ, 372, L71

Rauch, K.P., and Blandford, R.D. 1991, ApJ, 381, L39

Rauch, M., Carswell, R.F., Chaffee, F.H., Foltz, C.B., Webb, J.K., Weymann, R.J., Bechtold, J., and Green, R.F. 1992, ApJ, 390, 387

Rees, M.J. 1987, MNRAS, 228, 47P

Rees, M.J. 1989, MNRAS, 239, 1P

Rees, M.J., Silk, J.I., Werner, M.W., Wickramasinghe, N.C. 1969, Nature, 223, 788

Reichert, G.A., Mushotzky, R.F., Petre, R., and Holt, S.S. 1985, ApJ, 296, 69

Reichert, G.A., Polidan, R.S., Wu, C.-C., and Carone, T.E. 1988, ApJ, 325, 671

Reichert, G.A., et al. 1994, ApJ, 425, 582

Rieke, G.H. 1978, ApJ, 226, 550

Rieke, G.H., and Low, F.J. 1972, ApJ, 176, L95

Robertson, H.P. 1935, ApJ, 82, 284

Robinson, I., Schild, A., Schucking, E.L. 1965, Phys. Today, 18, No. 7, 17

Roos, N. 1985, ApJ, 294, 479

Ryle, M., and Sandage, A. 1964, ApJ, 139, 419

Salpeter, E.E. 1964, ApJ, 140, 796

Sandage, A. 1965, ApJ, 141, 1560

Sandage, A., and Luyten, W.J. 1967, ApJ, 148, 767

Sanders, D.B., Phinney, E.S., Neugebauer, G., Soifer, B.T., and Matthews, K. 1989, ApJ, 347, 29

Sargent, W.L.W. 1988, in *QSO Absorption Lines: Probing the Universe*, ed. J.C. Blades, D. Turnshek, and C.A. Norman (Cambridge University Press: Cambridge), p. 1

Sargent, W.L.W., Boksenberg, A., and Steidel, C.C. 1988, ApJS, 68, 539

Sargent, W.L.W., Young, P.J., Boksenberg, A., and Tytler, D. 1980, ApJS, 42, 41

Schechter, P. 1976, ApJ, 203, 297

Scheuer, P.A.G. 1965, Nature, 207, 963

Scheuer, P.A.G., and Readhead, A.C.S. 1979, Nature, 277, 182

Schmidt, M. 1963, Nature, 197, 1040

Schmidt, M. 1968, ApJ, 151, 393

Schmidt, M. 1969, in *Quasars and High-Energy Astronomy*, ed. K.N. Douglas, I. Robinson, A. Schild, E.L. Schucking, J.A. Wheeler, and N.J. Woolf (Gordon and Breach: New York), p. 55

Schmidt, M. 1970, ApJ, 162, 371

Schmidt, M., and Green, R.F. 1983, ApJ, 269, 352

Schmidt, M., Schneider, D.P., Gunn, J.E. 1991, in *The Space Distribution of Quasars*, ed. D. Crampton (Astronomical Society of the Pacific: San Francisco), p. 109

Schmidt, M., Schneider, D.P., Gunn, J.E. 1995, AJ, 110, 68

Schneider, D.P., et al. 1993, ApJS, 87, 45

Seyfert, C. 1943, ApJ, 97, 28

Shanks, T., and Boyle, B.J. 1994, MNRAS, 271, 753

Shields, G.A. 1978, Nature, 272, 706

Shields, J.C., and Ferland, G.J. 1993, ApJ, 402, 425

Shields, J.C., Ferland, G.J., and Peterson, B.M. 1995, ApJ, 441, 507

Shklovsky, I.S. 1964, Astron. Zh., 41, 801

Simkin, S.M., Su, H.J., and Schwarz, M.P. 1980, ApJ, 237, 404

Smith, E.P., and Heckman, T.M. 1989, ApJ, 341, 658

Smith, E.P., Heckman, T.M., Bothun, G.D., Romanishin, W., and Balick, B. 1986, ApJ, 306, 64

Smith, H.J., and Hoffleit, D. 1963, Nature, 198, 650

Smoot, G.F., et al. 1992, ApJ, 396, L1

Soifer, B.T., Houck, J.R., and Neugebauer, G. 1987, ARAA, 25, 187

Sołtan, A. 1982, MNRAS, 200, 115

Spinrad, H., Djorgovski, S., Marr, J., and Aguilar, L. 1985, PASP, 97, 932

Sprayberry, D., and Foltz, C.B. 1992, ApJ, 390, 390

Steidel, C.C. 1990, ApJS, 72, 1

Steidel, C.C., and Sargent, W.L.W. 1987, ApJ, 318, L11

Steidel, C.C., and Sargent, W.L.W. 1991, ApJ, 382, 433

Stengler-Larrea, E.A., et al. 1995, ApJ, 444, 64

Stirpe, G.M. 1991, A&A, 247, 3

Stocke, J.T., Morris, S.L., Gioia, I., Maccacaro, T., Schild, R.E., and Wolter, A. 1990, ApJ, 348, 141

Stocke, J.T., Morris, S.L., Gioia, I.M., Maccacaro, T., Schild, R., Wolter, A., Fleming, T.A., and Henry, J.P. 1991, ApJS, 76, 813

Stocke, J.T., Morris, S.L., Weymann, R.J., and Foltz, C.B. 1992, ApJ, 396, 487

Stocke, J.T., Perlman, E.S., Wurtz, R., and Morris, S.L. 1991, in *The Space Distribution of Quasars*, ed. D. Crampton (Astronomical Society of the Pacific: San Francisco), p. 218

Stockton, A. 1978, ApJ, 223, 747

Storchi-Bergmann, T., Wilson, A.S., and Baldwin, J.A. 1992, ApJ, 396, 45

Strittmatter, P.A., Serkowski, K., Carswell, R., Stein, W.A., Merrill, K.M., and Burbidge, E.M. 1972, ApJ, 175, L7

Strömgren, B. 1948, ApJ, 108, 242

Sun, W.-H., and Malkan, M.A. 1989, ApJ, 346, 68

Sutherland, R.S., Bicknell, G.V., and Dopita, M.A. 1993, ApJ, 414, 510

Svensson, R. 1987, MNRAS, 227, 403

Svensson, R. 1990, in *Physical Processes in Hot Cosmic Plasmas*, ed. W. Brinkmann, A.C. Fabian, and F. Giovanelli (Kluwer Academic Publishers: Dordrecht), p. 357

Tadhunter, C., and Tsvetanov, Z. 1989, Nature, 341, 422

Tanaka, Y., et al. 1995, Nature, 375, 659

Tananbaum, H., Avni, Y., Branduardi, G., Elvis, M., Fabbiano, G., Feigelson, E., Giacconi, R., Henry, J.P., Pye, J.P., Soltan, A., and Zamorani, G. 1979, ApJ, 234, L9

Terlevich, R., Tenorio-Tagle, G., Franco, J., and Melnick, J. 1992, MNRAS, 255, 713

Terrell, J. 1977, Am. J. Phys., 45, 869

Toomre, A., and Toomre, J. 1972, ApJ, 178, 623

Tran, H.D. 1995, ApJ, 440, 597

Tran, H.D., Miller, J.S., and Kay, L.E. 1992, ApJ, 397, 452

Treves, A., Maraschi, L., and Abramowicz, M. 1988, PASP, 100, 427

Turner, T.J., and Pounds, K.A. 1989, MNRAS, 240, 833

Turner, T.J., Weaver, K.A., Mushotzky, R.F., Holt, S.S., and Madejski, G.M. 1991, ApJ, 381, 85

Turnshek, D.A. 1988, in *QSO Absorption Lines: Probing the Universe*, ed. J.C. Blades, D. Turnshek, and C.A. Norman (Cambridge University Press: Cambridge), p. 17

Tytler, D. 1982, Nature, 298, 427

Tytler, D., and Fan, X.-M. 1992, ApJS, 79, 1

Tytler, D., and Fan, X.-M. 1994, ApJ, 424, L87

Ulrich, M.-H. 1988, MNRAS, 230, 121

Ulrich, M.-H. 1989, in *BL Lac Objects*, ed. L. Maraschi, T. Maccacaro, and M.-H. Ulrich (Springer-Verlag: Berlin), p. 45

Urry, C.M., Padovani, P., and Stickel, M. 1991, ApJ, 382, 501

Usher, P.D., and Mitchell, K.J. 1978, ApJ, 223, 1

van den Bergh, S. 1975, ApJ, 198, L1

Veilleux, S. 1991, ApJ, 369, 331

Veilleux, S., and Osterbrock, D.E. 1987, ApJS, 63, 295

Véron-Cetty, M.-P., and Véron, P. 1991, ESO Sci. Rep. 10

Véron-Cetty, M.-P., and Woltjer, L. 1990, A&A, 236, 69

Viegas-Aldrovandi, S.M., and Contini, M. 1989, ApJ, 339, 689

Vogel, S.N., Weymann, R., Rauch, M., and Hamilton, T. 1995, ApJ, 441, 162

Voit, G.M., Weymann, R.J., and Korista, K.T. 1993, ApJ, 413, 95

von Montigny, C., *et al.* 1995, ApJ, 440, 525

Walker, A.G. 1936, Proc. Lon. Math. Sci., 42, 90

Walker, M.F. 1968, ApJ, 151, 71

Wambsganss, J., Paczyński, B., and Schneider, P. 1990, ApJ, 358, L33

Wanders, I., Goad, M.R., Korista, K.T., Peterson, B.M., Horne, K., Ferland, G.J., Koratkar, A.P., Pogge, R.W., and Shields, J.C. 1995, ApJ, 453, L87

Ward, M.J., Wilson, A.S., Penston, M.V., Elvis, M., Maccacaro, T., and Tritton, K.P. 1978, ApJ, 223, 788

Warren, S.J., Hewett, P.C., Irwin, M.J., and Osmer, P.S. 1991, ApJS, 76, 1

Warren, S.J., Hewett, P.C., and Osmer, P.S. 1994, ApJ, 421, 412

Webb, J.R., Smith, A.G., Leacock, R.J., Fitzgibbons, G.L., Gombola, P.P., and Shepherd, D.W. 1988, AJ, 95, 374

Webster, R.L., Francis, P.J., Peterson, B.A., Drinkwater, M.J., and Masci, F.J. 1995, Nature, 375, 469

Weedman, D.W. 1976, QJRAS, 17, 227

Weedman, D.W. 1983, ApJ, 266, 479

Weedman, D.W. 1986, *Quasar Astronomy* (Cambridge University Press: Cambridge)

Weinberg, S. 1972, *Gravitation and Cosmology: Principles and Applications of the General Theory of Relativity* (John Wiley & Sons: New York)

Weymann, R.J., Boroson, T.A., Peterson, B.M., and Butcher, H.R. 1978, ApJ, 226, 603

Weymann, R.J., Carswell, R.F., and Smith, M.G. 1981, ARAA, 19, 41

Weymann, R.J., Morris, S.L., Foltz, C.B., and Hewett, P.C. 1991, ApJ, 373, 23

Weymann, R.J., Williams, R.E., Beaver, E.A., and Miller, J.S. 1977, ApJ, 213, 619

Whittle, M. 1985a, MNRAS, 213, 33

Whittle, M. 1985b, MNRAS, 216, 817

Whittle, M. 1992, ApJ, 387, 121

Wilkes, B.J. 1984, MNRAS, 207, 73

Wilkes, B.J., and Elvis, M. 1987, ApJ, 323, 243

Wilkes, B.J., Schmidt, G.D., Smith, P.S., Mathur, S., and McLeod, K.K. 1995, ApJ, 455, L13

Wills, B.J., Netzer, H., Brotherton, M.S., Han, M., Wills, D., Baldwin, J.A., Ferland, G.J., and Browne, I.W.A. 1992, ApJ, 410, 534

Wills, B.J., Netzer, H., and Wills, D. 1985, ApJ, 288, 94

Wills, D., and Lynds, R. 1978, ApJS, 36, 317

Wilson, A.S., and Colbert, E.J.M. 1995, ApJ, 438, 62

Wilson, A.S., Elvis, M., Lawrence, A., and Bland-Hawthorn, J. 1992, ApJ, 391, L75

Wilson, A.S., and Heckman, T.M. 1985, in *Astrophysics of Active Galaxies and Quasi-Stellar Objects*, ed. J.S. Miller (University Science Books: Mill Valley), p. 39

Wilson, A.S., and Tsvetanov, Z.I. 1994, AJ, 107, 1227

Wilson, A.S., Ward, M.J., and Haniff, C.A. 1988, ApJ, 334, 121

Wolfe, A.M., Turnshek, D.A., Smith, H.E., and Cohen, R.D. 1986, ApJS, 61, 249

Woltjer, L. 1959, ApJ, 130, 38

Woltjer, L. 1990, in *Active Galactic Nuclei*, ed. T.J.-L. Courvoisier and M. Mayor (Springer-Verlag: Berlin), p. 6

Worrall, D.M., and Wilkes, B.J. 1990, ApJ, 360, 396

Yee, H.K.C. 1980, ApJ, 241, 894

Yee, H.K.C. 1987, AJ, 94, 1461

Yee, H.K.C., and Green, R.F. 1984, ApJ, 280, 79

Yee, H.K.C., and Green, R.F. 1987, ApJ, 319, 28

Zdziarski, A., Ghisellini, G., George, I.M., Svensson, R., Fabian, A.C., and Done, C. 1990, ApJ, 363. L1

Zdziarski, A., Johnson, W.N., Done, C., Smith, D., and McNaron-Brown, K. 1995, ApJ, 438, L63

Zel'dovich, Ya.B., and Novikov, I.D. 1964, Sov. Phys. Dokl., 158, 811

Zheng, W., Kriss, G.A., Davidsen, A.F., Lee, G., Code, A.D., Bjorkman, K.S., Smith, P.S., Weistrop, D., Malkan, M.A., Baganoff, F.K., and Peterson, B.M. 1995a, ApJ, 444, 632

Zheng, W., Pérez, E., Grandi, S.A., and Penston, M.V. 1995b, AJ, 109, 2355

Bibliography

Allen, C.W. 1973, *Astrophysical Quantities* (Athlone Press: London)

Antonucci, R.R.J. 1993, 'Unified Models for Active Galactic Nuclei and Quasars', in *Annual Reviews of Astronomy and Astrophysics*, Vol. 31, p. 473

Begelman, M.C., Blandford, R.D., and Rees, M.J. 1984, 'Theory of Extragalactic Radio Sources', Rev. Mod. Phys., 56, 255

Bicknell, G.V., Dopita, M.A., and Quinn, P.J., editors. 1994, *The First Stromlo Symposium: The Physics of Active Galaxies*, ASP Conference Series, Vol. 54 (Astronomical Society of the Pacific: San Francisco)

Blades, J.C., Turnshek, D., and Norman, C.A., editors. 1988, *QSO Absorption Lines* (Cambridge University Press: Cambridge)

Blandford, R.D., Netzer, H., and Woltjer, L. 1990, *Active Galactic Nuclei*, ed. T.J.-L. Courvoisier and M. Mayor (Springer-Verlag: Berlin)

Bregman, J. 1990, 'Continuum Radiation from Active Galactic Nuclei', A&A Rev., 2, 125

Bridel, A.H., and Perley, R.A. 1984, 'Extragalactic Radio Jets', in *Annual Reviews of Astronomy and Astrophysics*, Vol. 22, p. 319

Burbidge, E.M. 1967, 'Quasi-Stellar Objects', in *Annual Reviews of Astronomy and Astrophysics*, Vol. 5, p. 399

Burbidge, G.R., and Burbidge, M. 1967, *Quasi-Stellar Objects* (W.H. Freeman: San Francisco)

Courvoisier, T.J.-L., and Blecha, A., editors. 1994, *Multi-Wavelength Continuum Emission of AGN* (Kluwer: Dordrecht)

Crampton, D., editor. 1991, *The Space Distribution of Quasars*, ASP Conference Series, Vol. 21 (Astronomical Society of the Pacific: San Francisco)

Davidson, K., and Netzer, H. 1979, 'The Emission Lines of Quasars and Similar Objects', Rev. Mod. Phys., 51, 715

Douglas, K.N., Robinson, I., Schild, A., Schucking, E.L., Wheeler, J.A., and Woolf, N.J., editors. 1969, *Quasars and High-Energy Astronomy* (Gordon and Breach: New York)

Duschl, W.J., and Wagner, S.J., editors. 1992, *Physics of Active Galactic Nuclei* (Springer-Verlag: Berlin)

Duschl, W.J., Wagner, S.J., and Camenzind, M., editors. 1991, *Variability of Active Galaxies* (Springer-Verlag: Berlin)

Dyson, J.E., editor. 1985, *Active Galactic Nuclei* (Manchester University Press: Manchester)

Filippenko, A.V., editor. 1992, *Relationships Between Active Galactic Nuclei and Starburst Galaxies*, ASP Conference Series, Vol. 31 (Astronomical Society of the Pacific: San Francisco)

Frank, J., King, A.R., and Raine, D.J. 1985, *Accretion Power in Astrophysics*, 2nd Edition (Cambridge University Press: Cambridge)

Giuricin, G., Mardirossian, F., Mezzetti, M., and Ramella, M., editors. 1986, *Structure and Evolution of Active Galactic Nuclei* (Reidel: Dordrecht)

Gondhalekar, P.M., editor. 1987, *Emission Lines in Active Galactic Nuclei*, Rutherford Appleton Laboratory Report RAL-87-109 (Rutherford Appleton Laboratory: Chilton)

Gondhalekar, P.M., Horne, K., and Peterson, B.M., editors. 1994, *Reverberation Mapping of the Broad-Line Region in Active Galactic Nuclei*, ASP Conference Series, Vol. 69 (Astronomical Society of the Pacific: San Francisco)

Gunn, J.E., Longair, M.S., and Rees, M.J. 1978, *Observational Cosmology*, ed. A. Maeder, L. Martinet, and G. Tammann (Geneva Observatory: Geneva)

Hartwick, F.D.A., and Schade, D. 1990, 'The Space Distribution of Quasars', in *Annual Reviews of Astronomy and Astrophysics*, Vol. 28, p. 437

Hazard, C., and Mitton, S., editors. 1979, *Active Galactic Nuclei* (Cambridge University Press: Cambridge)

Kafatos, M., editor. 1988, *Supermassive Black Holes* (Cambridge University Press: Cambridge)

Kellermann, K.I., and Pauliny-Toth, I.I.K. 1981, 'Compact Radio Sources', in *Annual Reviews of Astronomy and Astrophysics*, Vol. 19, p. 373

Khachikian, E.Ye., Fricke, K.J., and Melnick, J., editors. 1987, *Observational Evidence of Activity in Galaxies* (Reidel: Dordrecht)

Kinman, T.D. 1975, 'Variable Quasi-Stellar Sources with Particular Emphasis on Objects of the BL Lac Type', in *Variable Stars and Stellar Evolution*, ed. V.E. Sherwood and L. Plaut (Reidel: Dordrecht), p. 573

Kormendy, J., and Richstone, D. 1995, 'Inward Bound – The Search for Supermassive Black Holes in Galactic Nuclei', in *Annual Reviews of Astronomy and Astrophysics*, Vol. 33, p. 581

Lang, K.R. 1974, *Astrophysical Formulae* (Springer-Verlag: Berlin)

Lang, K.R. 1992, *Astrophysical Data: Planets and Stars* (Springer-Verlag: Berlin)

Maraschi, L., Maccacaro, T., and Ulrich, M.-H., editors. 1989, *BL Lac Objects* (Springer-Verlag: Berlin)

Miller, H.R., and Wiita, P.J., editors. 1988, *Active Galactic Nuclei* (Springer-Verlag: Berlin)

Miller, H.R., and Wiita, P.J., editors. 1991, *Variability of Active Galactic Nuclei* (Cambridge University Press: Cambridge)

Miller, J.S., editor. 1985, *Astrophysics of Active Galaxies and Quasi-Stellar Objects* (University Science Books: Mill Valley)

Mushotzky, R.F., Done, C., and Pounds, K.A. 1993, 'X-Ray Spectra and Time Variability of Active Galactic Nuclei', in *Annual Reviews of Astronomy and Astrophysics*, Vol. 31, p. 717

Novikov, I.D., and Thorne, K.S. 1973, 'Astrophysics of Black Holes', in *Black Holes*, ed. C. DeWitt and B.S. DeWitt (Gordon and Breach: New York), p. 343

Osmer, P.S., Porter, A.C., Green, R.F., and Foltz, C.B., editors. 1988, *Proceedings of a Workshop on Optical Surveys for Quasars*, ASP Conference Series, Vol. 2 (Astronomical Society of the Pacific: San Francisco)

Osterbrock, D.E. 1984, 'Active Galactic Nuclei', QJRAS, 25, 1

Osterbrock, D.E. 1989, *Astrophysics of Gaseous Nebulae and Active Galactic Nuclei* (University Science Books: Mill Valley)

Osterbrock, D.E. 1991, 'Active Galactic Nuclei', Rep. Prog. Phys., 54, 579

Osterbrock, D.E. 1993, 'The Nature and Structure of Active Galactic Nuclei', ApJ, 404, 551

Osterbrock, D.E., and Mathews, W.G. 1986, 'Emission-Line Regions of Active Galaxies and QSOs', in *Annual Reviews of Astronomy and Astrophysics*, Vol. 24, p. 171

Osterbrock, D.E., and Miller, J.S., editors. 1989, *Active Galactic Nuclei* (Kluwer: Dordrecht)

Pacholczyk, A.G. 1970, *Radio Astrophysics* (W.H. Freeman: San Francisco)

Peebles, P.J.E. 1971, *Physical Cosmology* (Princeton University Press: Princeton)

Peebles, P.J.E. 1993, *Principles of Physical Cosmology* (Princeton University Press: Princeton)

Peterson, B.M. 1988, 'Emission-Line Variability in Seyfert Galaxies', PASP, 100, 18

Peterson, B.M. 1993, 'Reverberation Mapping of Active Galactic Nuclei', PASP, 105, 247

Pringle, J.E. 1981, 'Accretion Disks in Astrophysics', in *Annual Reviews of Astronomy and Astrophysics*, Vol. 19, p. 137

Rees, M.J. 1984, 'Black Hole Models for Active Galactic Nuclei', in *Annual Reviews of Astronomy and Astrophysics*, Vol. 22, p. 471

Robinson, A., and Terlevich, R., editors. 1994, *The Nature of Compact Objects in Active Galactic Nuclei* (Cambridge University Press: Cambridge)

Robinson, I., Schild, A., and Schucking, E.L. 1965, *Quasi-Stellar Sources and Gravitational Collapse* (University of Chicago Press: Chicago)

Rybicki, G.B, and Lightman, A.P. 1979, *Radiative Processes in Astrophysics* (John Wiley and Sons: New York)

Schmidt, M. 1975, 'Quasars', in *Stars and Stellar Systems, Vol. 9. Galaxies and the Universe*, ed. A. Sandage, M. Sandage, and J. Kristian (University of Chicago Press: Chicago), p. 283

Shlosman, I., editor. 1994, *Mass-Transfer Induced Activity in Galaxies* (Cambridge University Press: Cambridge)

Svensson, R. 1990, 'An Introduction to Relativistic Plasmas in Astrophysics', in *Physical Processes in Hot Cosmic Plasmas* (Kluwer Academic Publishers: Dordrecht), p. 357

Swarup, G., and Kapahi, V.K., editors. 1986, *Quasars* (Reidel: Dordrecht)

Ulfbeck, O., editor. 1978, 'Quasars and Active Nuclei of Galaxies', Physica Scripta, Vol. 17, No. 3

Urry, C.M., and Padovani, P. 1995, 'Unified Schemes for Radio-Loud Active Galactic Nuclei', PASP, 107, 803

Valtaoja, E., and Valtonen, M., editors. 1992, *Variability of Blazars* (Cambridge University Press: Cambridge)

Verschuur, G.L., and Kellermann, K.I., editors. 1988, *Galactic and Extragalactic Radio Astronomy* (Springer-Verlag: Berlin)

Wagner, S.J., and Witzel, A. 1995, 'Intraday Variability in Quasars and BL Lac Objects', in *Annual Reviews of Astronomy and Astrophysics*, Vol. 33, p. 581

Ward, M.J., editor. 1994, *Proceedings of the Oxford Torus Workshop*, privately distributed

Weedman, D.W. 1977, 'Seyfert Galaxies', in *Annual Reviews of Astronomy and Astrophysics*, Vol. 15, p. 69

Weedman, D.W. 1986, *Quasar Astronomy* (Cambridge University Press: Cambridge)

Weinberg, S. 1972, *Gravitation and Cosmology: Principles and Applications of the General Theory of Relativity* (John Wiley & Sons: New York)

Weymann, R.J., Swihart, T.L., Williams, R.E., Cocke, W.J., Pacholczyk, A.G., and Felten, J.E. 1976, *Lecture Notes on Introductory Theoretical Astrophysics* (Pachart: Tucson)

Wolfe, A.M., editor. 1978, *Pittsburgh Conference on BL Lac Objects* (University of Pittsburgh: Pittsburgh)

Index